FUNDAMENTALS OF

FAULT CURRENT AND GROUNDING IN ELECTRICAL SYSTEMS

HALDEN MORRIS & NORMAN CHAMBERS

Order this book online at www.trafford.com
or email orders@trafford.com

Most Trafford titles are also available at major online book retailers.

Printed in the United States of America.

ISBN: 978-1-4907-3561-0 (sc)
ISBN: 978-1-4907-3562-7 (hc)
ISBN: 978-1-4907-3563-4 (e)

Library of Congress Control Number: 2014908589

Trafford rev. 06/30/2014

www.trafford.com

North America & international
toll-free: 1 888 232 4444 (USA & Canada)
fax: 812 355 4082

CONTENTS

LIST OF TABLES

PREFACE

The book, Fundamentals of Fault Current and Grounding in Electrical Systems was written as a result of extensive, controversial conversations on grounding electrode, grounding configurations, and fault current. Despite upgraded and safe methods of grounding techniques, the misconceptions on grounding and fault current are indelible in the minds of many engineers, technicians, and contemporaries.

This book seeks to explain in simple terms the behavior of fault current through the general mass of earth, the origin of short-circuit current and its value, and how a circuit breaker operates. The drawings are unique and allow the reader to visualize the behavior of a fault current. The book clarifies common myths pertaining to a grounding electrode, short circuit, and open neutral conditions, and provides an unambiguous understanding of the theoretical and practical explanation for an effective earthing and protective system in all electrical installations.

There are numerous grounding problems and unexplained fault conditions in electrical circuitry that are taken for granted and left unattended for extended periods. Potential voltage can be found on the earthing conductors in processing plants, refineries, and other small industrial plants. Many of these potentials are due to induction, which could be due to a number of motors in use at the same time, creating a magnetic field great enough to induce a potential in nearby earth conductors. A combination of topics in this book addresses this problem and other likely problems, which have been adversely affecting the electrical industry for many years. Introduction of ground fault circuit interrupt (GFCI)systems, ungrounded systems, clean earthing system, motor control systems, resistance grounding systems, lightning protection systems, and intra earthing systems are included in the book as mechanisms employed in providing solutions to some of these problems.

Many sources of information were consulted when writing this book. Sources include the *IEE On-site Guide*, the IEE 17th edition, the Canadian Electrical Code, the National Electrical Code (NEC), *Newnes Electrical Pocket Book*, *Marks' Standard Handbook for Mechanical Engineers*, GE application information, *Fluke 434 Analyzer User Manual*, *Schneider Electric HV Training Manual*, *Square D Manual*, and other manuals, journals, and Web sites mentioned in the references at the end of each chapter. The information and all the codes used as references are common to most countries, even though they may differ with respect to peculiarities such as the color coding of wires and systems of measure. Furthermore, research has shown that despite language barriers, basic electrical practices are common throughout the world.

The target audience and topics covered in this book fall within the scope of most electrical personnel, which include, but is not limited to, emergent engineers, electrical technicians, technical vocational education and training (TVET) practitioners, and students who need a working overview of the behavior of short-circuit current in order to develop a working knowledge of the level of short-circuit current available at an installation, and the origin of fault current. This knowledge also helps to clarify the supposed association between the damage that is done to circuits and appliances due to open neutral conditions and the consumers' earth electrode. A Test Your Knowledge section is provided at the end of each chapter in order to facilitate students.

Morris and Chambers

ACKNOWLEDGMENT

We acknowledge and extend our heartfelt gratitude to the following persons: Mrs. Carlene Morris, Ms. Tracey-Ann Burnett, Mr. Kirkland Lawrence, Mr. Hugh Sandford, Mr. Paul DaCosta-Pinto, and Mr. Solomon Burchell for their vital encouragement and technical support. Sincere thanks to our families for their understanding and support. Thanks to Almighty God, who gives life, strength, wisdom and purpose.

ABBREVIATIONS

a.c.	Alternating Current
AFCI	Arc Fault Circuit Interrupter
AIC	Asymmetrical Interrupting Current
CPC	Circuit Protective Conductor
CPR	Cardiopulmonary Resuscitation
CT	Current Transformer
d.c.	Direct current
ELCB	Earth Leakage Circuit Breaker
EMT	Electrical Metal Tube
GEI	Government Electrical Inspector
GFCI	Ground Fault Circuit Interrupter
HSE	Health and Safety Executive
HRC	High Rupturing Capacity
HV	High Voltage
Hz	Hertz
IDMTL	Inverse Definite Minimum Time Log
ICU	Ultimate or final short-circuit breaking capacity
ICS	Service short-circuit breaking capacity
IEE	Institute of Electrical Engineers
JS31	Jamaican Standard 31st Edition
kA	Kilo Ampere
kWH	Kilo Watt Hour
LPS	Lightning Protection System
LRA	Lock Rotor Ampere
LV	Low Voltage
MDP	Main Distribution Panel
MOSFET	Metal Oxide Semiconductor Field Effect Transistor
NEC	National Electrical Code

NEDOB	Neutral, Earth, Dissipation, and opening of Circuit Breaker
OCD	Overcurrent Devices
OCPD	Overcurrent Protective Device
IGBT	Insulated Gate Bipolar Transistors
IOT	Input Output Terminal
PT	Potential Transformer
PAS	Parallel Arc Signature
%Z	Percentage Impedance
PVC	Polyvinyl Chloride
PFC	Prospective Fault Current
QTL	Quick Trip Link
RCD	Residual Current Device
r.m.s.	Root Mean Square
SAS	Series Arc Signature
SCPD	Special Circuit Protection Device
SCR	Silicon Control Rectifier
SMPS	Switch Mode Power Supply
THD	Total Harmonic Distortion
TN	Earth and Neutral Combined
TT	Installation of Earth Electrode and Source Earth Electrode (Multiple Earthing System)
TRIAC	Triode Alternating Current Switch)
UPS	Uninterrupted Power Supply
USA	United States of America
UL	Underwriters laboratories
VFD	Variable Frequency Drive
VT	Voltage Transformer
X_L	Inductive reactance
Z_S	Earth Loop Impedance

Information used from the *National Electrical Code* in this book has been included with permission accordingly:

CHAPTER 1

The Fundamentals of Electrical Circuits

1.0 Introduction

In order to get an indepth understanding of earthing systems, it is important that a review of the fundamentals of electrical circuits be done. This is the foundation from which knowledge of advanced studies in electrical technology and engineering is built. In this chapter, the simple circuit, circuit protection, safety, and basic electrical concepts will be explored.

1.1 Simple Electric Circuit

An electrical circuit is a combination of four elements connected in a single loop. Figure 1.1 shows a simple circuit, which comprises a voltage source, a load, and a switch connected in a series configuration with conducting wires.

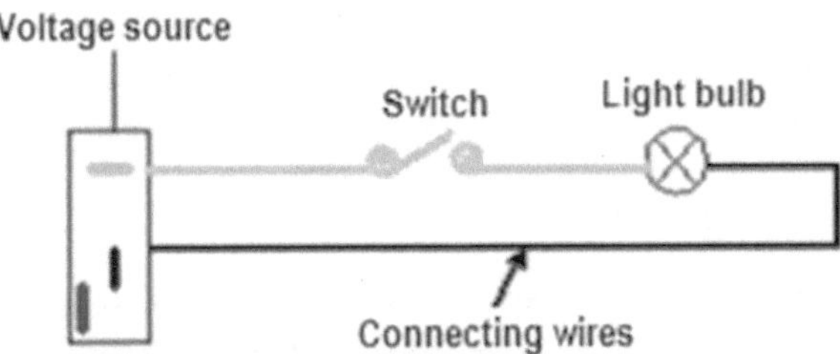

Figure 1.1: *Simple electric circuit*

Complex devices are sometimes employed to enhance safety in electric circuits. Devices such as overcurrent protective devices (OCPD) are used to ensure that the current through the circuit elements does not exceed predetermined parameters. Overcurrent devices (OCDs) are designed to quickly isolate the circuit devices/appliances from their voltage sources if the current exceeds the required values. A circuit may exceed its predetermined parameters as a result of an overload or a short circuit.

1.2 Circuit Protection

Local electrical codes require that all electrical installations shall have adequate mechanical or insulation protection, grounding protection, and overcurrent protection. Circuit protection requires that unwanted currents and voltages be removed from electrical circuits as quickly as possible. Fuses, circuit breakers, and relays provide critical protection by disconnecting unwanted current and voltage within fraction of a second to avoid damage to the circuit components.

1.2.1 Mechanical Protection

Mechanical protection prevents direct access by unintended mechanical force to cables, circuit connections, and devices, and reduces the risk of physical damage to these elements. As shown in Figure 1.2, mechanical protection can be provided by using polyvinyl chloride (PVC) or metal enclosures, PVC or electrical metal tube (EMT)/rigid conduits, or any other provision which will provide adequate mechanical protection for cable and devices in an electrical installation.

In general, conduits used in an electrical installation may be exclusively EMT, exclusively PVC, or a combination of both EMT and PVC. In all cases where conduits are used, insulated conductors are pulled through the conduits and terminated in junction boxes.

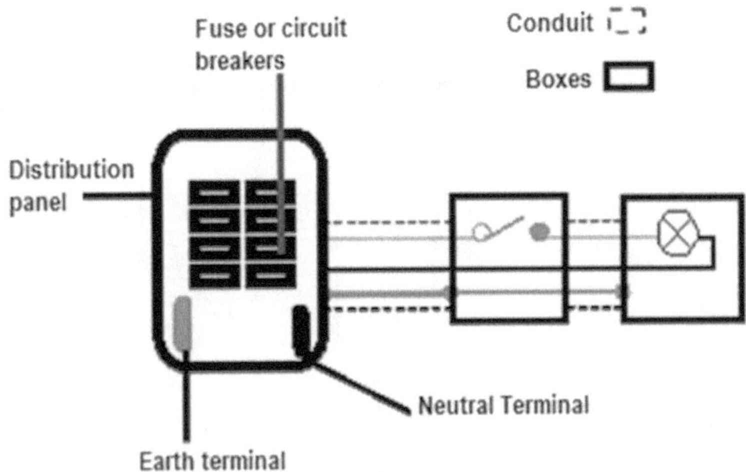

Figure 1.2: *Basic layout of panels, conduits, and boxes*

1.2.2 Conduit Installation

Conduits play an extremely vital role in all electrical installations. They protect cables from environmental and mechanical damages. Electrical conduits can be overloaded by virtue of the number of cables being pulled through them. The term *diversity factor* refers to the degree to which a conduit is being loaded.

Overloading of conduits will cause overheating of conductors, which can lead to melting of cable insulation. Once the insulation of the cables inside an overloaded conduit is damaged, a short circuit is likely to occur, which will result in the loss of all cables inside that conduit. If all cables are damaged, it may require total rewiring and redesigning of the installation before returning it to service.

Figure 1.3 shows the layout of conduits, panel box, receptacle boxes, and outlet boxes for an installation. The method of conduiting shown in this figure provides guidance on basic conduiting works which separate circuits from each other and make wiring simple and easy. This method of conduiting reduces the risk of total electrical failure of an installation. If a fire occurs on any circuit, it will be limited to the circuits which are within that conduit.

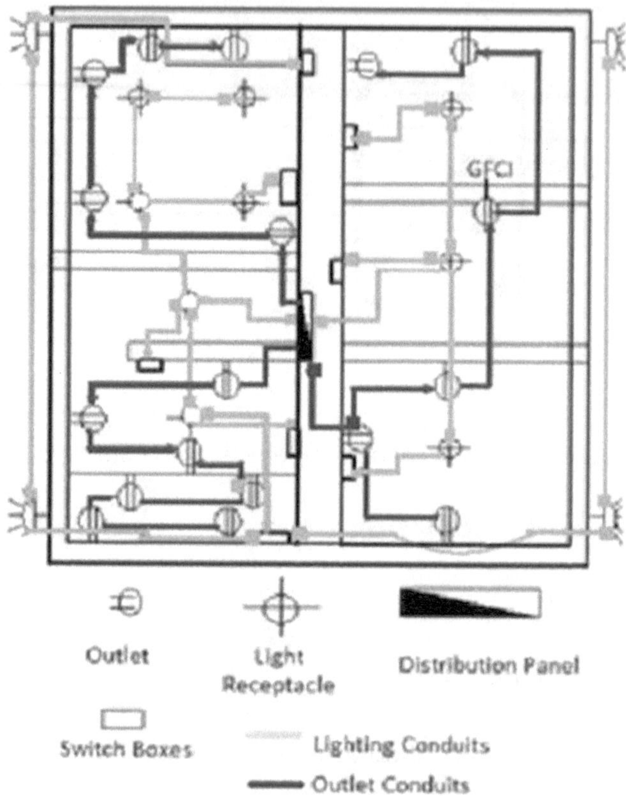

Figure 1.3: *Typical electrical conduit layout*

1.2.2.1 Bending Conduits

Bending electrical conduits, according to specification, eliminates the waste of conduits and damage of cables during pulling. Table 1.1 provides guidance on bending specification in accordance with the following criteria:

1. Radius of conduit bends
2. Size of conduits
3. Tolerance for the bend

Table 1.1: Electrical conduit sizes and radius of bend

Conduit Size In Inches	Conduit Size In MM	Tolerance Inches to MM	Radius of Bend Inches to MM
½	16	1.50 – 38.1	4.00 – 102
¾	21	1.50 – 38.1	4.50 – 114.3
1	27	1.875 – 47.6	5.75 – 146
1-1/4	35	2.00 – 50.8	7.25 – 184
1-1/2	41	2.00 – 50.8	8.25 – 209.6
2	53	2.00 – 50.8	9.50 – 241.3
2-1/2	63	3.00 – 76.2	10.50 – 226.7
3	78	3.125 – 79.38	13.00 – 330.2
3-1/2	91	3.25 – 82.55	15.00 – 381
4	103	3.375 – 85.73	16.00 – 406.4
5	129	31.70 – 94	24.00 – 609.6
6	155	31.70 - 94	36.00 – 914.4

Where T = Tolerance, R = Radius of bend
(Adapted from Scepter Rigid PVC Conduit and Fittings, 2009)

Bending of conduits requires using precise measurements to form exact radii and fit. Incorrect radii can lead to the kinking of the conduits. Once a conduit is deformed or kinked, the internal radius is decreased, and pulling of cables may result in damage to the cable insulation.

1.2.2.2 Bending of Rigid Conduits

Figures 1.4 shows the correct procedure for bending a ½ in rigid conduit, using the bending tolerance from Table 1.1. It is imperative that the bending tolerance for various size conduits be observed since this differ for each size. Failing to recognize this will result in wastage of material and improper bends.

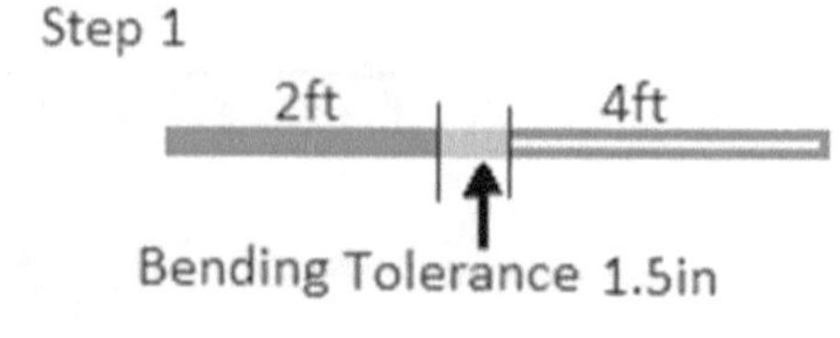

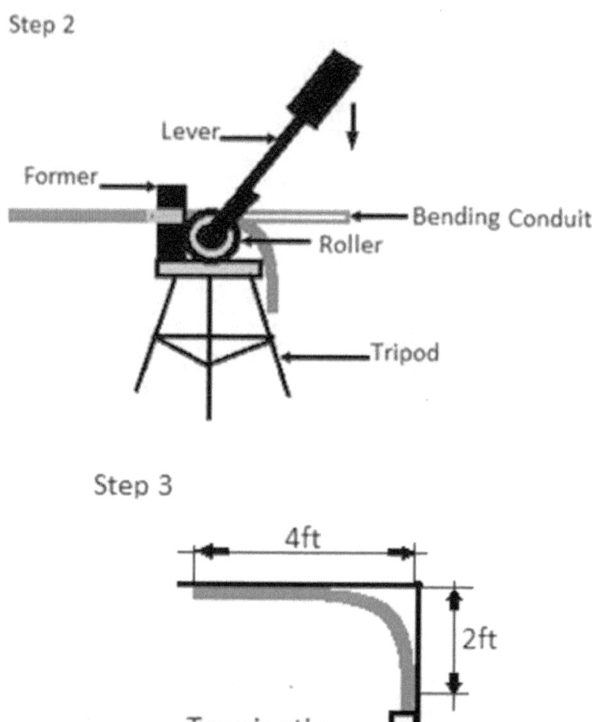

Figure 1.4: *Bending rigid electrical conduits using a tripod or hickey*

1.2.2.3 Bending of PVC Conduits

Bending of PVC conduits can be achieved by using an electric heater blower, a portable propane torch, spring benders, or PVC heating blankets among many other methods. To avoid deformation of the conduit while bending, conduits and bending equipment must be used in accordance with the manufactures' specifications. Figures 1.5: A and B, Figures 1.6 A and B, and Figures 1.7: A and B show various methods of bending PVC conduits.

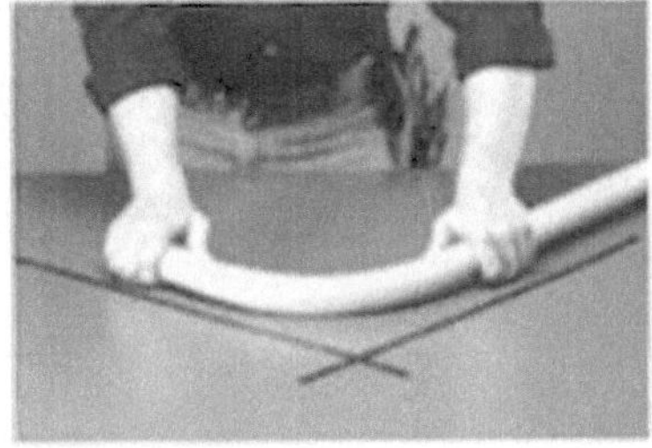

(A) (B)

(Adapted from Scepter Rigid PVC Conduit and Fittings, 2009)

Figure 1.5: *Preparing to bend a PVC electrical conduit using an electric heater or propane torch as shown in (A)*

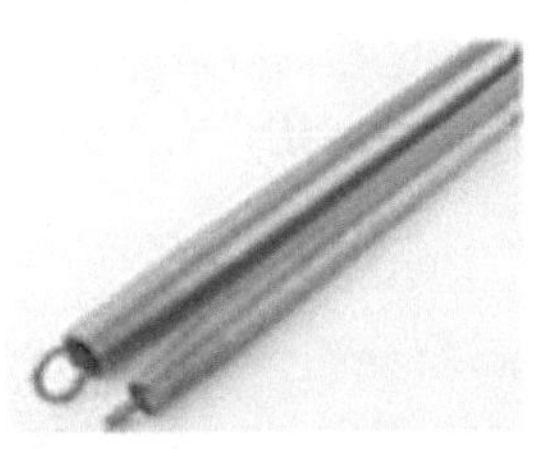

(A) Spring Bender (B) Bending of Conduit

(Adapted from Google images for using a spring bender for PVC conduit)

Figure 1.6: *Bending PVC electrical conduits with the use of a spring bender*

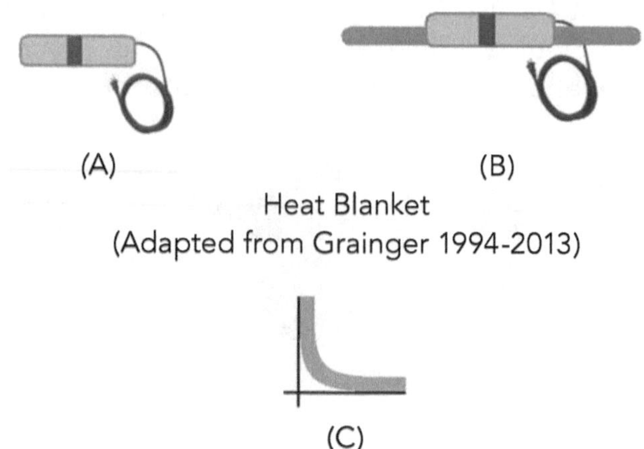

Figure 1.7: *Bending PVC electrical conduits to 90° with the use of a heating blanket*

1.2.3 Ground Protection

Grounding is an integral part of all electrical installation. It provides the correct path for fault current to return to earth through the circuit's earth terminal. A fault is generated if a live conductor unintentionally comes in contact with an exposed grounded metal part of an installation.

Ground protection is simply connecting the metal frame of appliances or machineries to the main neutral bar and the general mass of earth to reduce the risk of electrocution. *If an individual comes in contact with a live, exposed conductive part on an appliance, for example, refrigerator, washing machine, or dryer, the following effects can occur to the body in a matter of seconds if the current flowing exceeds 4 mA.*

1. Negligible sensation or shock.
2. Finger muscles contract and fail to give up their grip almost immediately.
3. Restriction of breathing begins.

4. Disorientation of the control signals to the heart. When this occurs, the heart may be forced to stop working and may result in death.

It is, therefore, imperative that all electrical equipment are properly maintained to protect life and property by ensuring that a solid and effective grounding system exists at all times. Grounding and bonding connections reduce the hazards posed by fault current with the potential of causing electrocution. Figure 1.8 shows typical grounding and neutral connections inside a distribution panel.

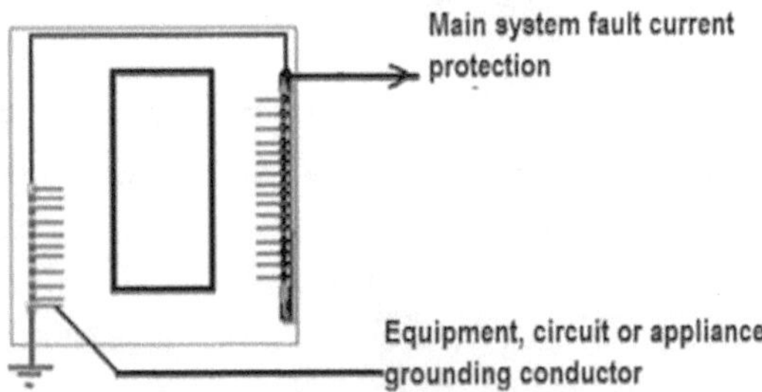

Figure 1.8: *Typical connection for circuit, equipment, and appliance protection at the distribution panel*

1.2.4 Short-circuit Protection

Circuits are protected from damage due to short-circuit situations by the use of devices such as circuit breakers and fuses, which are rated below the rating of the cable and at the rating of the devices being protected.

If a short circuit is created inside a metal switch box as shown in Figure 1.9, the circuit breaker will activate and isolate the circuit immediately from the voltage source whenever the switch is turned on.

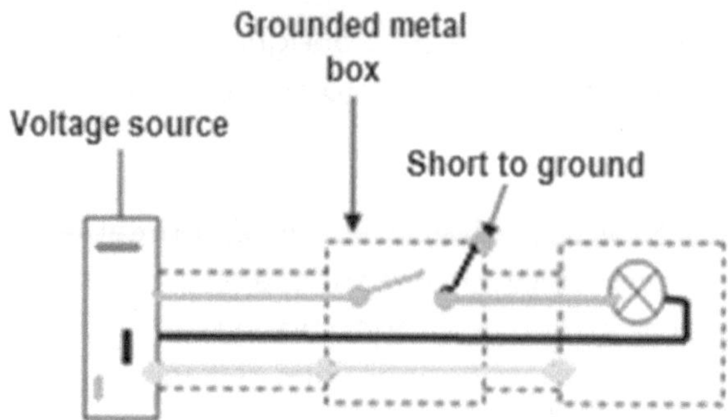

Figure 1.9: *Short circuit created inside a switch box*

1.2.5 Overcurrent Protection

Circuits are protected from overcurrent situations by the use of circuit breakers, fuses, and other current protection devices. When a load exceeds the capacity of a circuit, the circuit protection device will isolate the load from the voltage supply. For example, when a circuit with a 3.5 kW halogen bulb is activated in a circuit rated at 20 A, as shown in Figure 1.10, the protection device will isolate that device momentarily since the current drawn by this device is above the capacity of the circuit protection device.

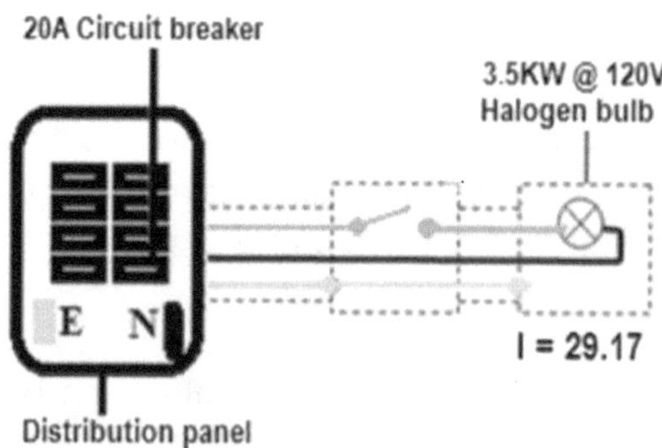

Figure 1.10: *Overcurrent due to a high-wattage bulb*

1.3 Practical Electrical Circuit

Most practical electrical circuits contain a combination of loads and multiple protection devices, which may include the following:

1. OCPDs
2. Main earthing conductor
3. Main neutral conductor
4. Equipment/circuit earthing conductor
5. Earth rod/electrode
6. Earthing clamp
7. Earthing bar in the supply panel
8. The general mass of earth

A practical electrical circuit which incorporates several protective elements and loads is shown in Figure 1.11.

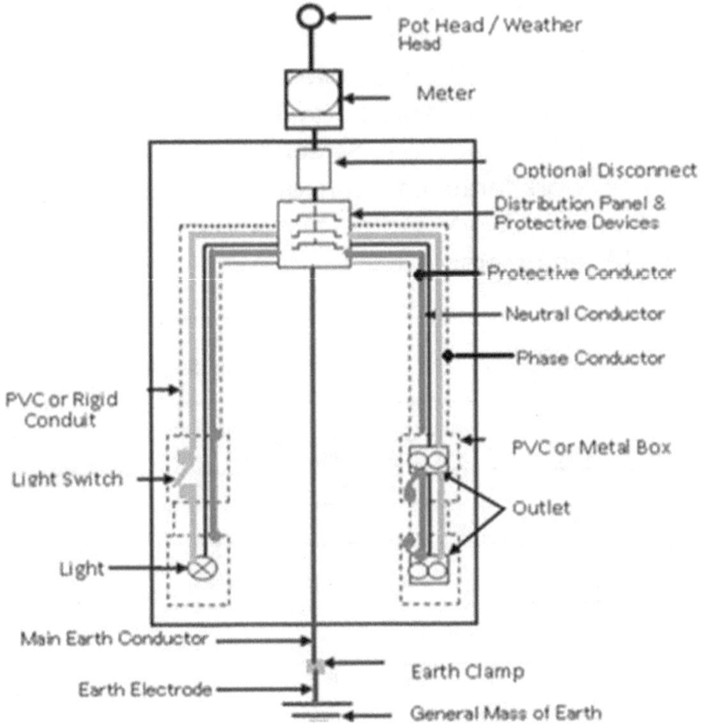

Figure 1.11: *Typical electrical installation with circuit protection*

1.4 Safe Work Practices

Safety considerations are critical when performing a task which involves interaction with electricity. Safety procedures are usually associated with these tasks, and deviation from such procedures can result in malfunctioning equipment, damage to property, serious injuries, or even loss of life.

Measures should be taken to mitigate against personal injuries, loss of life, or damage to property during electrical work activities or during the use of electrical equipment or appliances. These safety measures are not limited to but are guided by a set of electrical codes such as the JS31, the National Electrical Code (NEC), the IEE, or the Canadian Electricity Code, which were established by consultation and adhering to best practices.

Electrical injuries are caused by and include, but is not limited to, electric shock, electric burns, electrical explosions or arcing, as well as electrical fires. There are several procedures for working safely with electricity as depicted by Health and Safety Executive (HSE) in the *Electricity at Work, Safety and You, 2003***.** In assessing whether safe work practices can be achieved in areas or on equipment on which work is to be carried out, the flowchart shown in Figure 1.12 can be employed.

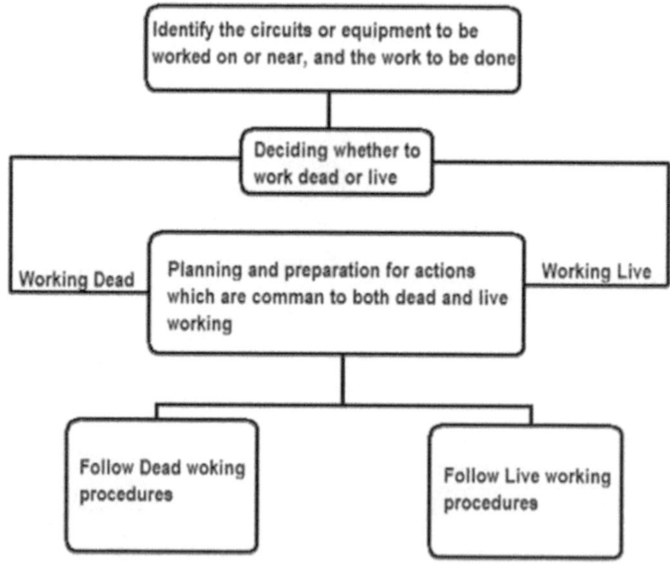

(Adapted from *HSE Electricity Safety and You*, 2003)

Figure 1.12: *Assessing safe work practices*

The Electricity at Work Act outlined procedures in Figure 1.13 in the form of a flowchart for deciding whether to work on dead or live circuits. Figure 1.14 further outlines the procedures for working dead, and Figure 1.15 provides procedures for working live.

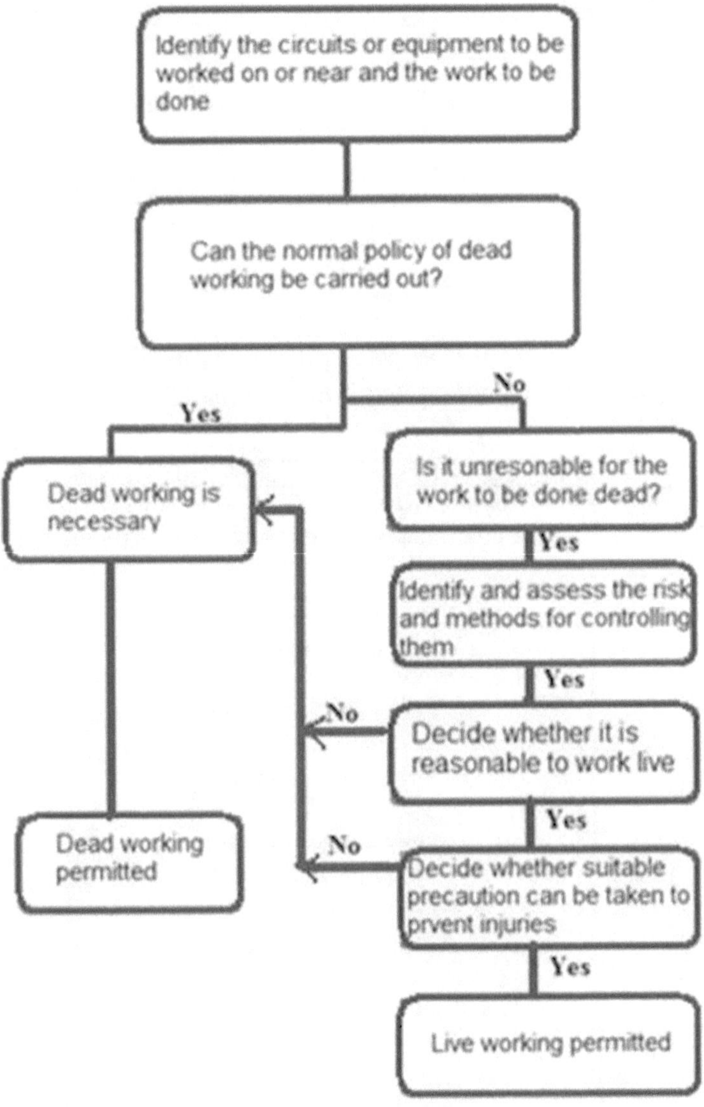

(Adapted from *HSE Electricity Safety and You*, 2003)

Figure 1.13: *Flow chart to decide whether to work on dead or live circuits*

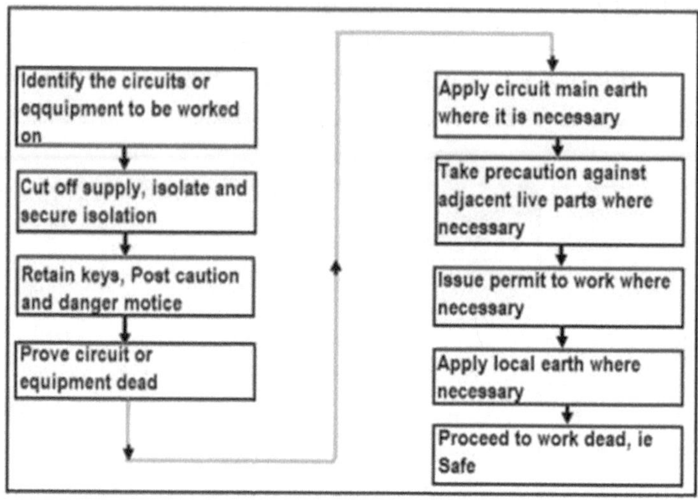

(Adapted from *HSE Electricity Safety and You*, 2003)
Figure 1.14: *Deciding when to work dead*

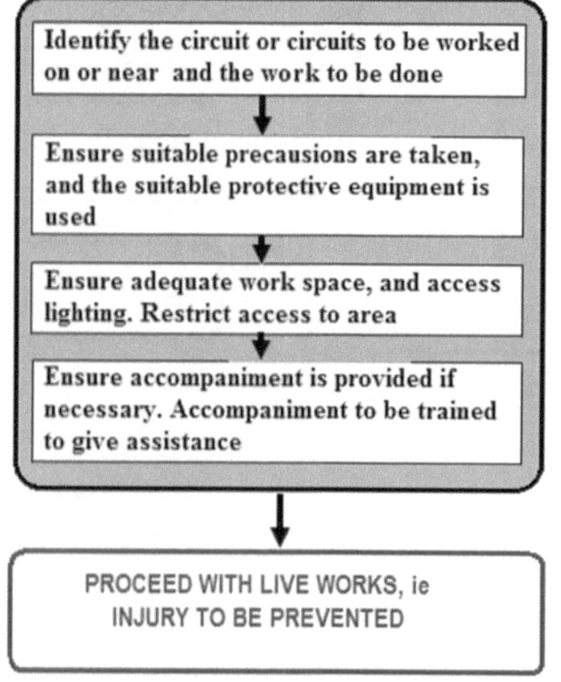

(Adapted from *HSE Electricity Safety and You*, 2003)
Figure 1.15: *Deciding when to work live*

1.5 Troubleshooting Electrical Installations

Trouble shooting an electrical installation involves probing the circuitry to identify faults. This process can be made simple if the principles of operation regarding installations are understood and followed. The following are the three installation protection characteristics to be kept in mind:

1. Current protection (overcurrent device)
2. Touch protection (grounding protection)
3. Insulation protection (mechanical protection)

Most electrical protective devices are triggered by faults associated with the type of protective device protecting the circuit. It is for this reason that some electrical faults are easily identified.

1.5.1 Electrical Faults

Electrical faults are abnormal electrical occurrences which will cause a monitoring or protective device to operate and isolate the voltage source. Faults such as those resulting from moisture, phase-to-ground short, phase to neutral short, phase to phase short, low voltage (LV) and high voltages (HV) are common.

Presented in Table 1.2 are additional types of electrical faults which may result in loss of power, intermittent tripping of circuit breaker, electrical shock, and frequent damage to circuit devices and loads.

Table 1.2: Typical electrical faults

Loss of Power	Intermittent Tripping	Electric Shock	Damage to Load
Blown fuse, Broken conductor(s), Corrosion, ONC, Tripped overload, Tripped circuit breaker, loose terminals	Circuit overload, Defective circuit breaker, Defective electronics, Defective equipment, Defective motors, Mechanical defects, No neutral-to-ground bonding	Defective grounding, Defective appliance, Defective equipment, Direct contact with a live conductor, Short to ground Improper wiring	Loose connection, Water in junction boxes, Existence of moisture, Over voltage, Loss of phase

1.6 Additional Safety Tips

1. Familiarity breeds contempt; treat all routine tasks as if it were a first approach.
2. Incompetence compromises safety; never attempt to carry out a task without being competent.
3. Be extraordinarily calm and meticulous when working live or dead.
4. Always use insulated tools.
5. Use the correct test instruments to carry out the type of test you are attempting to do.
6. Identify nearest emergency exits.
7. Never attempt to work in a wet area on live circuits.
8. Always wear electrical gloves, simple latex for dead work and testing and UL-rated (Underwriters Laboratories) HV gloves for live work.
9. Wear electrical insulated boots when working on electrical and electronic circuits.

10. Wear electrical insulated gloves when working on electrical and electronic circuits.
11. Wear a helmet when on the job.
12. Wear eye protection when required.
13. Always use electrical tools such as pliers and screwdrivers with their insulation rated at 1000 V.
14. Implement lockout and tag-out systems where possible.

1.7 Conclusion

Whether the installation is small or large, the principle of installation remains the same. A simple circuit comprises a voltage source, a load, and a switch connected in a series configuration with connecting conducting wires. Mechanical protection is essential for all electrical cables. Protection may take the form of PVC or EMT conduits, enclosures, and protective boxes.

It is imperative that safe work practices are exercised to prevent nicking of conductors, which causes short circuits, fire, and electrocution. Keen attention must be given to the loading and the sizing of conduit in an effort to provide free air space when cables are drawn through them. Free air space in conduits prevents cables from exceeding normal operating temperatures. Free space also allows the easy pulling of cables and reduces the risk of harmful rubbing of cable against the conduit. In electrical construction and maintenance, good practices will eliminate the fear of damages or injuries and will promote a safe working environment.

1.8 Test Your Knowledge:

1. What is an electrical circuit?
2. Draw and label a simple electrical circuit.
3. Name the basic components of an electrical circuit.
4. What is the purpose of an earth conductor in an installation?
5. What effect would a fault current have on the user of an appliance if a fault current is present and the grounding system is defective?
6. Explain the difference between a short circuit and an overload.
7. State three methods of providing mechanical protection for electrical cables.
8. State three types of protection that an electrical installation requires.
9. Why is it important to ground all metal boxes in an electrical installation?
10. What steps are to be taken before attempting to engage in an electrical task?
11. What is electrical safety?
12. List three methods used for monitoring potential fire hazards.
13. List four fire alarm initiating devices.
14. In a fire alarm system, what is the device used to signal persons with sight and hearing disabilities?
15. What is the key feature of an electrical tool?
16. What is the primary cause of an electric shock?
17. List four primary causes for loss of power.
18. List five primary causes of intermittent tripping of circuit protection devices.

Bibliography

i. Donnelly, E. L. 1985. *Electrical Installation Theory and Practice* Pages 54-67. Published 1985 by Thomas Nelson and Sons Ltd. Nelson house, Mayfield Road, Walton-on-Thomas Surry, KT12 5PL UK
ii. Jamaica Bureau of Standard. 1992. Jamaican standard specification for electrical installations.
iii. National Fire Protection Association. 2008. National Electrical Code. National Fire Protection Association [Section 250-64].
iv. NFPA International. 2002. National Fire Alarm Code. NFPA 72.
v. NFPA International. 2002. National Fire Alarm Code. NFPA 13.
vi. NT WorkSafe. 2001. First Aid in the Workplace. Retrieved on April 30th, 2014, from http://www.worksafe.nt.gov.au/Publications/Code%20of%20Practice/first_aid_cop.pdf
vii. Google images for using a spring bender for PVC conduit https://www.google.com/search?q=images+for+using+a+spring+bender+for+PVC+conduit&tbm=isch&tbo=u&source=univ&sa=X&ei=dBjiUcTjEcL84APukoHwCw&ved=0CCwQsAQ&biw=1366&bih=667
viii. HSE. Electrical Safety and You. Retrieved on January 16, 2010, from http://www.hse.gov.uk/pubns/indg231.pdf http://www.hse.gov.uk/firstaid/legislation.htm
ix. Workcover NSW. 2001. First Aid in the Workplace. Retrieved on August 26, 2012, from http://redcross.e3learning.com.au/content/legal/NSW.pdf

CHAPTER 2

Codes for Electrical Installation

2.0 Introduction

Electrical codes are a set of standards established to guide safe installation of electrical accessories, cables, and equipment. These codes are prescribed by law, and all installations shall meet the minimum standards stipulated by the codes.

Electrical installations incorrectly done can pose a high risk of electrocution or serious injuries. Electrical work practices must, therefore, follow the established electrical codes. To assure safety, all electrical installations are inspected by government electrical inspectors or GEIs. This activity is established to eliminate poor work practices carried out by electricians and contractors. Electrical inspections are done prior to connections to the utility's power supply or as stipulated by a manufacturer of the device(s) being installed.

2.1 The Mandate of an Electrical Code or Regulation

The electrical codes or regulations are guidelines documented to provide protection of persons, livestock and property, from damage which may occur because of unethical electrical work practices during the electrical installation process. Each code or regulation is guided by the auxiliary *shall*. The auxiliary *shall* in any electrical code indicates that the point of reference is mandated by law. This means that deviation from the code of reference may present serious consequences if the matter is taken to court for matters arising from breaches.

The auxiliary *May* is conscionable and is left to one's discretion about the point of reference. Strict attention must be paid to these key words; these terms, along with their references, are your only protection in a court of law.

An electrical code is not a stand-alone manual and provides minimum standards for maximum safety. Therefore, other regulations *may* be required as is necessary to supplement the electrical code where additional equipment or apparatuses are used and are not a part of the normal electrical installation. For example, fire detection and suppression systems or electric signs and HV luminaries *may* require additional codes.

2.2 Minimum Code Requirements

Since the objective of electrical codes is to provide a guide for the protection of persons, livestock, and property, the minimum protection for any electrical installation shall be within the following six parameters:

1. Protection against direct contact
   (Using current-limiting circuit breakers)
2. Protection against indirect contact
   (Using barriers)
3. Protection against thermal effect
   (Placing barriers between hazardous or flammable materials keeping them from electrical systems)
4. Protection against overcurrent
   (Using circuit breakers and current-limiting devices)
5. Protection against fault current
   (Provides adequate bonding of all metal parts within the installation to the main earth and neutral terminals)
6. Protection against overvoltage
   (Provides voltage-monitoring devices where necessary)

In addition to the six critical installation criteria, electrical codes provide guidelines for the installation of mechanical protection for all cables, access boxes, and accessory boxes. There are accessories, apparatuses, and equipment which are not covered in the electrical code. These are limitations to the electrical code to which additional information or references should be sought.

All electrical installations must be done in accordance with established codes to enhance safety and to facilitate the following:

1. The purpose of the installation (fuel dispenser, factory, domestic, etc.)
2. The type of supply or service required (three-phase or single-phase and type of voltage)
3. The environment (corrosive atmosphere, wet, extremely hot, etc.)
4. The type of protective devices (current-limiting, GFCI, regular thermal device, etc.)
5. The type of wiring method (surface or flush)
6. The type of mechanical protection (PVC or rigid conduit, duct, etc.)
7. The type of accessories to be fitted (flame proof or regular)

2.3 Code for Circuit Loading

The code below gives guidance on how circuits should be safely loaded. When circuits are correctly loaded, heat dissipation is minimized on the circuit conductors. The NEC stipulates permissible loads in Article 210-23 as follows:

> In no case shall the load exceed the branch circuit ampere rating. An individual branch circuit shall be permitted to supply any load for which it is rated. A branch circuit supplying two or more outlets or receptacles shall supply only the loads specified according to its size as specified in (a) through (d) . . .

A 15 or 20 ampere branch circuit shall be permitted to supply lighting units or other utilization equipment" *such as A/C units,* "or a combination of both. The rating of any one cord and plug connected utilization equipment, or fastened in place shall not exceed 80 percent of the branch circuit ampere rating. The total rating of utilization equipment such as an A/C unit fastened in place shall not exceed 50 percent of the branch circuit ampere rating where lighting, units cord and plug connected utilization equipment and not fastened in place, or both are also supplied.

To illustrate this, see Figure 2.1.

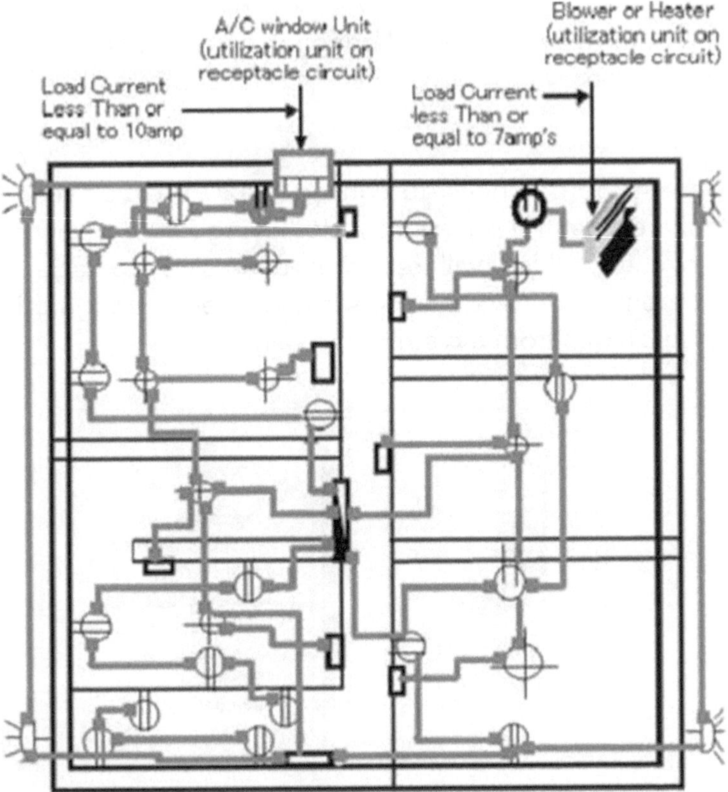

Figure 2.1: *Addition of equipment (blower, heater, etc.) to a lighting or outlet circuit*

2.4 Sharing of Neutral between Circuits

Every single-phase 120 V circuit consists of one live conductor and one neutral conductor. This basic electrical theory shall never be compromised. The meaning of this theory is: if five (5) single pole circuit breakers are located inside a distribution panel, there should be five (5) neutral conductors connected to the neutral bar corresponding to the five single pole circuit breakers located inside the same panel. Sharing neutral conductors to reduce cost or time are shortcuts that can have devastating effects on life or property. The most serious risk associated with this practice is being electrocuted during troubleshooting. It is, therefore, imperative that this condition does not exist in an electrical installation.

Another risk which involves sharing of neutral between circuits involves the removal of power from a critical piece of equipment unknowingly.

In the event there is need to convert a 120 V circuit to a 240 V circuit, shared neutral will result in deploying 220 V to unintended 110 V circuits, which will result in extensive damage to 110 V equipment. Figure 2.2 shows the correct configuration for deploying circuits from a distribution panel.

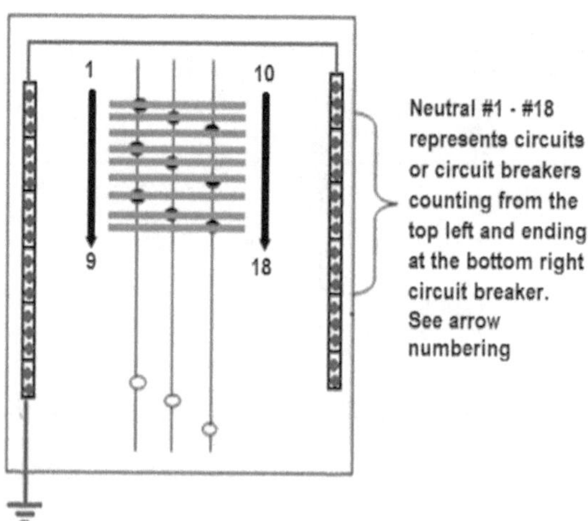

Figure 2.2: *Deploying circuits from distribution panel*

2.5 Branch Circuits

Article 100 of the NEC stipulates the types of branch circuits as follows:

1. Branch circuit—Appliance
2. Branch circuit—General purpose
3. Branch circuit—Individual

2.5.1 Branch Circuit—Appliance

An appliance branch circuit is a circuit that supplies power to one or more outlets to which appliances are to be connected and which has no permanent connection made to lighting fixtures, which are not a part of the connected appliance.

Permanent connection to fixtures means lighting fixtures or other similar fixtures permanently fixed or connected to other lighting fixture or lamp holders. Typical appliance branch circuits are shown in Figure 2.1.

2.5.2 Branch Circuit—General Purpose

A general purpose branch circuit is a circuit that supplies a number of outlets for lighting and appliances as shown in Figure 2.3.

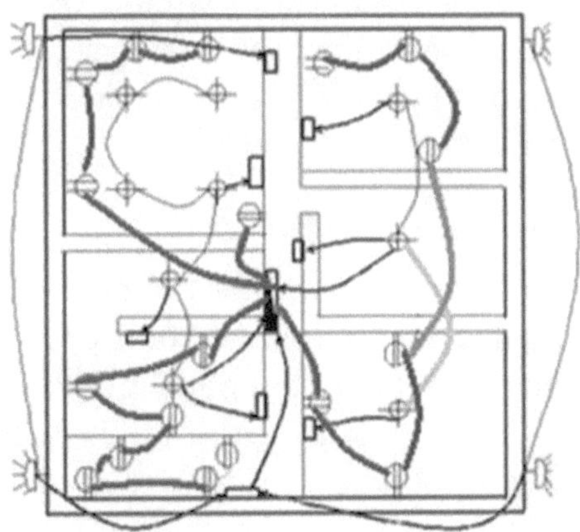

Figure 2.3: *Branch circuits—general purpose*

2.5.3 Branch Circuit—Individual

An individual branch is a circuit that supplies single utilization equipment only such as a cooker as shown in Figure 2.4.

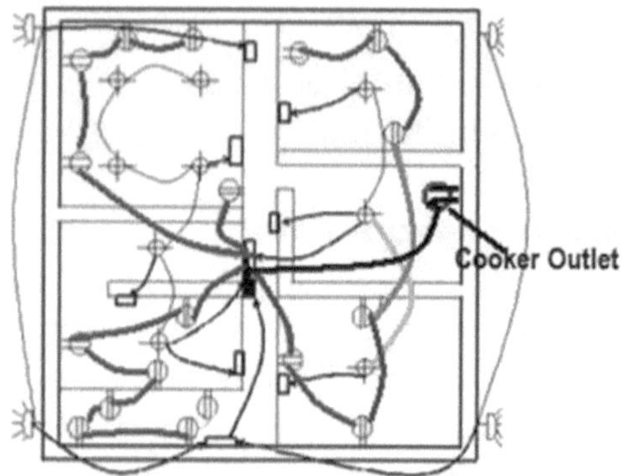

Figure 2.4: *Individual branch circuit (cooker outlet)*

2.6 Panel Board and Cable Protection

All panel boards shall be protected by a set of disconnects or circuit breaker protective devices as shown in Figure 2.5. The following Article 408.36 (a) from the NEC dictates panel board protection:

> Each lighting and appliance branch-circuit panel board shall be individually protected on the supply side by not more than two main circuit breakers or two sets of fuses having a combined rating, not greater than that of the panel board.
>
> Exception No. 1: Individual protection for a lighting and appliance panel board shall not be required if the panel board feeder has over-current protection, not greater than the rating of the panel board.

Where a single or combination meter center is used, and where the meter center does not incorporate individual disconnect or circuit breaker, the main disconnect or circuit breaker located inside the panels will suffice. Where meter facility facilitates disconnects or circuit breakers, disconnects or main circuit breaker shall be fitted on both ends of the main cable, supplying a main distribution panel (MDP).

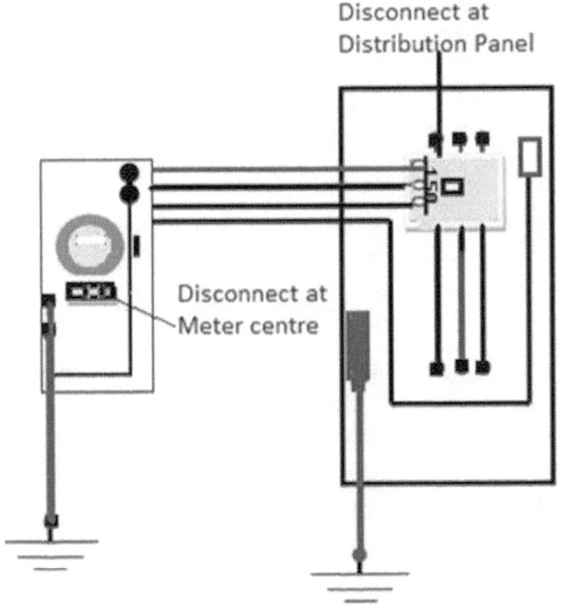

***Figure 2.5**: Disconnect at meter facility and distribution panel*

2.7 Fixed or Stationary Appliances

All fixed appliances shall be fitted with a supplementary isolation/ disconnecting device. The following sections 462-01-01, 463-01-01, and 464-01-01 of the IEE wiring regulation dictate the use of a functional switch or disconnecting devices.

Section 462-01-01: Switching off for mechanical maintenance

> A means of switching off for mechanical maintenance shall be provided where mechanical maintenance may involve a risk of burns or a risk injury from mechanical movement.

Section 463-01-01: Emergency switching

> A means of emergency switching shall be provided for every part of an installation which may be necessary to isolate a circuit rapidly from the supply in order to prevent or remove danger.

Section 464-01-01: Functional switching (control)

> A functional switching device shall be provided for each part of a circuit which may be required to be controlled independently of other parts of the installation.

The following Article 422-31 (a and b) of the NEC provides guidance on the use of disconnecting means for *Fixed or Stationary Appliances*.

Article 422-31 (a):

> For permanently connected appliances rated at, not over 300 VA or 1/8 horsepower, the branch circuit over-current device shall be permitted to serve as the disconnect.

Article 422-31 (b):

> For permanently connected appliances rated over 300 VA or 1/8 horsepower, the branch circuit switch or circuit breaker shall be permitted to serve as the disconnecting means where the switch or circuit breaker is within sight from the appliance or is capable of being locked in the open position.

In reference to Article 422-31 (a) and (b) of the NEC, 300 VA or 1/8 horsepower means any appliance with current rating below 1 A. The code also makes reference to "within sight"; this reference should not be confused with "within reach." A circuit breaker can be within sight but not within reach. All fixed appliances should be fitted with a "means of disconnect." The safety switches *shall* be located within 6 feet or 2.5 m of the appliance being supplied with power as shown in Figure 2.6.

In general, consumers have a fear of electricity and, therefore, demonstrate hesitance in touching an electrical distribution panel. However, most consumers are comfortable touching an electrical switch or pressing an electrical control button to turn off a light or to stop a machine. It is, therefore, extremely essential for *functional switches* and disconnects *switches* to be installed for fixed appliances (stoves, dryers, waters heaters, and other dedicated power appliance) in close proximity to the appliances and within reach of the user.

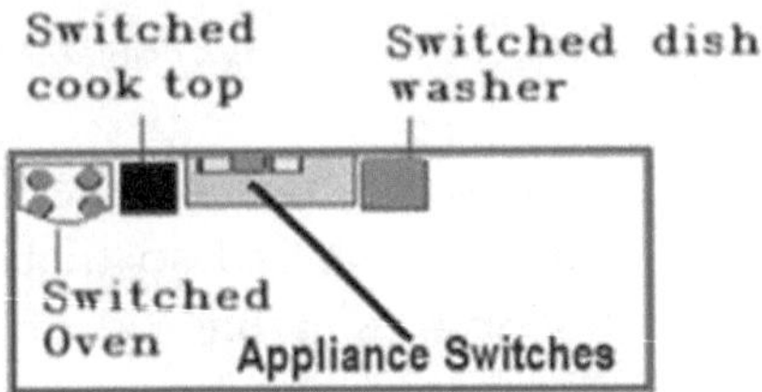

Figure 2.6: *Layout of fixed appliances and their disconnects*

2.8 Panel Board Accessibility

All buildings with interconnecting levels, for example, a block of offices with each set of offices occupying two or more levels or floors, shall be fitted with an *accessibility* distribution board on each level as shown in Figure 2.7.

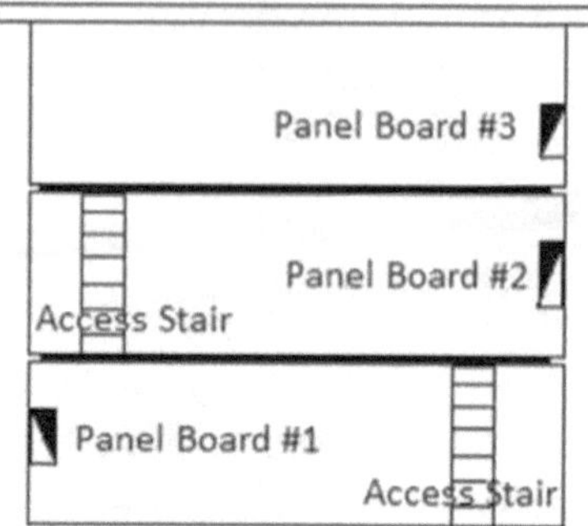

Figure 2.7: *Accessibility distribution panel on each level of building*

Article 240-24 (a) of the NEC defines accessibility:

> Accessibility: Over-current devices shall be readily accessible and shall be installed so that the center of the grip of the operating handle of the switch or circuit breaker when in its highest position, is not more than 2.0 m (6 feet-7 in) above the floor or working platform unless one of the following applies: For busways, as provided in Section 364-12 . . .

Article 100 of the NEC defines "readily accessible" as capable of being reached quickly for operation, renewal, or inspection, without requiring those, to whom ready access is requisite, to climb over or remove obstacles or to resort to portable ladders. In many instances, this code is compromised when panel boards are covered with paintings or pictures.

2.9 Safety Switches

Safety switches or disconnects shall be installed on all fixed appliances and must at all times be visible as shown in Figure 2.8. The purpose of safety switches and distribution panels is sometimes compromised when they are locked away in locations where they are difficult to access in the event of an emergency.

Figure 2.8: *Accessible disconnect for A/C unit*

Electrical safety apparatus should be placed in locations where they are easily visible and accessible. During the construction stage of an installation, electrical drawings should be reviewed for safety and breaches of the electrical code. Where breaches of the electrical code are evident, corrective action should be taken immediately.

2.10 Motor Protection

All motors rated more than 1 A shall be fitted with a device for overload protection. One may argue that the circuit breaker connected to the motor provides adequate protection for the connected load. This would be true if the starting current of motors was the same as the line current and the running current. The starting current of a motor is significantly higher than its continuous running current because of the characteristics of the magnetic coils.

Circuit breakers protect the motor from tripping when starting and also protect the cable from overloading. The windings of a motor can handle the starting current required to put the rotor in motion. Since the running current is significantly less than the starting current, an overload protection rated above the running current is required for all motors. The NEC and IEE codes for protection of electric motors stated below dictate the requirements for overload protection. Figure 2.9 shows the layout of such protection.

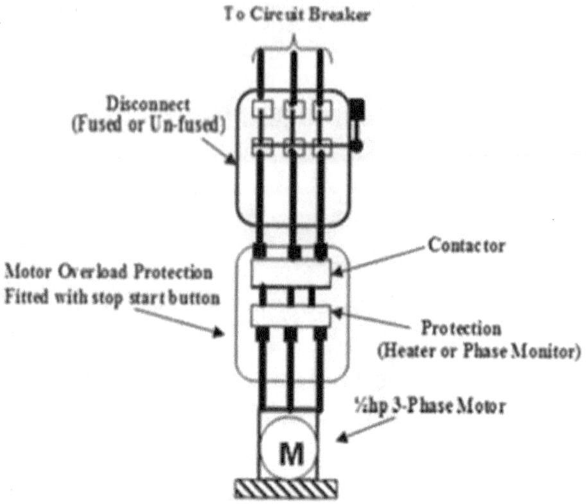

Figure 2.9: *Typical motor safety protection*

Article 430-32 of the NEC provides guidance on the protection of electric motors as shown in Figure 2.9 as follows:

> Each continuous-duty motor application rated more than 1 hp shall be protected against overload by one of the following means: (1) A separate overload device that is responsive to motor current. This device shall be selected to trip or shall be rated at no more than the following percent of the motor nameplate full-load current rating; (2) A thermal protector integral with the motor, approved for use with the motor it protects on the basis that it will prevent dangerous over heating of the motor due to overload and failure to start.

Section 552-01 from the IEE 16th Edition also provides guidance on the protection of electric motors and is as follows:

> Every electric motor having a rating exceeding 0.37 kW shall be provided with control equipment incorporated means of protection against overload of the motor.

2.10.1 Lock Rotor Current

An induction motor produces a starting force which is created by the magnetic field of the motor. This starting force is known as the *starting torque*, which is measured in pounds per foot or Newton meter. Starting torque creates a turning motion of the motor shaft from rest to full speed of the motor and reduces significantly when the shaft is fully in motion. Conversely, starting torque can produce a fault current known as *lock rotor current* if it is sustained on a motor over a long period of time. The graph in Figure 2.10 shows the relationship between delta and Y/Star torque and their associated currents.

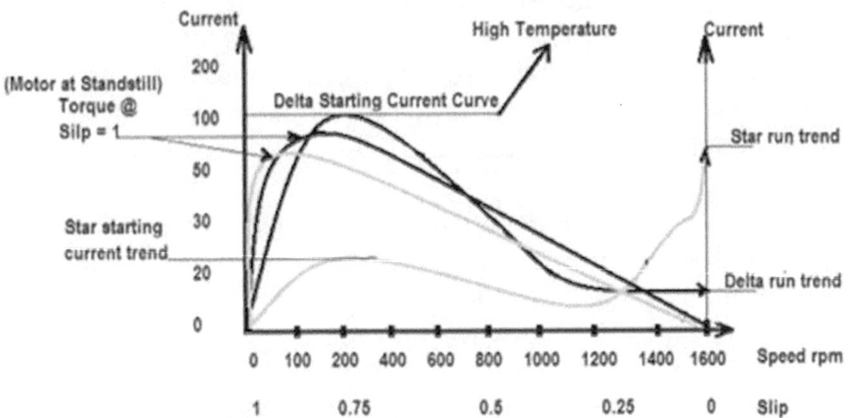

Figure 2.10: *Delta torque current versus star torque current*

Lock rotor current is created by a motor when the speed of the rotor of a motor is suddenly reduced while power is being supplied to that motor. Motor speed can be significantly reduced by the jamming of a pump shaft, ceased or damaged rotor bearings, or any other similar mechanical fault.

All motors have a design letter code. This letter code is found on the motor specification plate, which represents the power in kVA per horsepower. In many instances, starting current or lock rotor current is assumed to be 300% or three times the normal running current of the motor. In some cases, the lock rotor current may be as high as eight times the normal running current.

Motor maintenance is critical in all industries. It is imperative that the starting current of motors be evaluated to determine the condition of a motor prior to placing it in operation. Higher than normal starting current could signify potential failure of bearings or other connected equipment. Starting current should never be taken for granted. It is essential that code letters are understood and factored in calculations to determine the starting current or the lock rotor current of motors.

The NEC Articles 430.7-B-2 and 4 (single speed motors) dictate the requirements for the use of the motor code letters. This code gives specification for different types of motors and must be applied to determine the actual lock rotor current of motors. Table 2.1 gives guidance for motor code letters.

> Single-speed motors: starting on "Wye" connection and running on delta connections shall be marked with the letter code corresponding to the locked-rotor Kilovolt-ampere (kVA) per horsepower.
>
> 50/60 Hz Motor: Motors with 50 and 60 Hz ratings shall be marked with a code letter designating the locked-rotor kilovolt-ampere (kVA) per horsepower on 60 Hz.

The formula to calculate lock-rotor current (I_{lr}) is as follows:

$$\mathbf{I}_{\mathbf{lr}} = \frac{\text{hp} \times \text{letter code} \times 1000}{\text{Voltage} \times 1.732}$$

Table 2.1: Lock rotor indicating code letters

Code Letter	Kilovolt-Amperes per Horsepower with Locked Rotor
A	0-3.14
B	3.15-3.54
C	3.55-3.99
D	4.0-4.49
E	4.5-4.99
F	5.0-5.59
G	5.6-6.29
H	6.3-7.09
J	7.1-7.99
K	8.0-8.99
L	9.0-9.99
M	10.0-11.19
N	11.2-12.49
P	12.5-13.99
R	14.0-15.99
S	16.0-17.99
T	18.0-19.99
U	20.0-22.39
V	22.4 and up

Example 2.1:

A motor to be installed is rated at 15 Hp 3 phase, and the motor has a letter code of K. What will be the starting current of that motor with an operating voltage of 208 V? The corresponding kVA for the letter code K is 8.99.

$$\textit{Starting Current}\,(I) = \frac{15 \text{ x } 8.99 \text{ x } 1000}{208 \text{ x } 1.732}$$

$$= \frac{134850}{360.25}$$

$$374.58$$

Example 2.2:

Using Example 2.1, determine the running current of this motor.

$$\textbf{\textit{Running Current}}(I) = \frac{\text{hp x 746}}{\text{Voltage x 1.732}}$$

$$= \frac{\text{15 x 746}}{\text{208 x 1.732}}$$

$$= \frac{11190}{360.25} = 31.1A$$

In Examples 2.1 and 2.2, the calculation shows the starting current of the motor to be 374.58 A and the running current to be 31.1 A. Should the running current alone be used to select the overload protection for a motor, constant tripping may occur while the motor attempts to start. If the LRA is used to select the overload, there will be a catastrophic failure of the cable and motor insulation. One significant factor remaining to be considered when selecting motor overload is the service factor marked (SF) on all motor specification plate as shown in Figure 2.11.

A service factor signifies the percentage at which a motor is designed to operate above its normal running current. Furthermore, if a service factor of the motor in Example 2.2 is 1.5, it means that the motor can operate no more than 50% above its rated current of 31.1 A; overload protection of this motor should not exceed 31.1 A × 1.5 = 46.65 A. In motors where letter code is 1, a service factor of 1.15 should be used.

The control of current acting on copper conductors inside a motor is a crucial part of electrical engineering. Circuit breakers and fuses do not offer adequate protection to motors while in operation. A motor relies significantly on monitoring devices and protective relays for their protection.

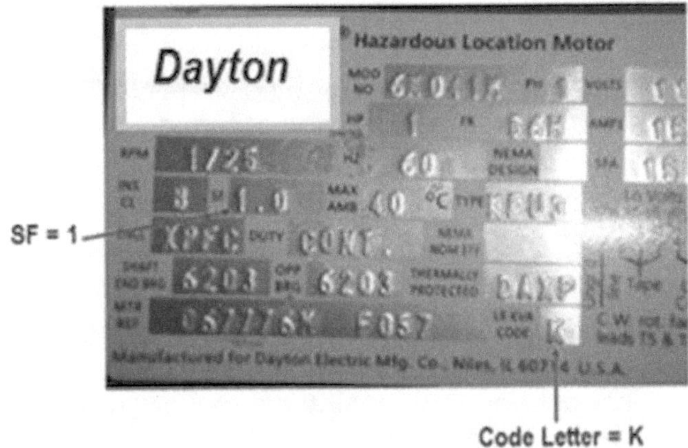

Figure 2.11: *Specification plate showing service factor (SF) and code letter K*

2.11 Transformers

Pad-mounted utility distribution transformers shall be given appropriate protection against damage by external forces. These transformers shall also be protected against risk they may pose to life and property. One significant risk a transformer poses to life and property is rapturing of the oil casing, which may or may not be flammable. Transformers may also be exposed to high risks, depending on where they are located, such as from flying debris or motor vehicle collision. Despite the potential risk which surrounds domestic pad-mounted transformers, they are sometimes used inappropriately for purposes such as *seating or storage for objects.*

The NEC Article 450-8, Guarding of Transformers (Pad-mounted), gives guidance on the protection of transformer.

> Transformers shall be guarded as specified as follows:
>
> Mechanical Protection: Appropriate provisions shall be made to minimize the possibility of damage to transformers from external causes where the transformers are exposed to physical damage.

Case or Enclosed: Dry-type transformers shall be protected with a non-combustible, moisture-resistant case and enclosure that will provide protection against the accidental insertion of foreign objects.

The NEC Article 450-13 gives guidance to the accessibility of transformers.

All transformers and transformer vaults shall be readily accessible to qualified personnel for inspection and maintenance or meet the requirements of (a) or (b).

a) Open installation: Dry transformers 600 Volts nominal, or less, located in the open on walls, columns, or structures, shall not be required to be readily accessible.
b) "Hollow Space Installation: Dry-type transformers 600 Volts, nominal or less and not exceeding 50 kVA, shall be permitted in hollow spaces of buildings not permanently closed in by structure, and provided they meet the ventilation requirements of section 459-9, and separation from combustible materials requirements on section 450-21(a). Transformers so installed shall not be required to be readily accessible.

In addition to Articles 450-13 (a) and (b), it is extremely crucial to be aware of the importance of barriers for pad-mounted transformers where they are used. Barriers as shown in Figure 2.12 are necessary and mandatory because they reduce the risk of serious injury, loss of life, and damage to property.

Figure 2.12: *Transformer barrier*

2.12 Conclusion

Electricity can be very lethal and destructive if electrical installations are incorrectly done or if a body unintentionally comes in contact with a live conductor. It is for this reason electrical work practices are closely monitored and electrical codes are readily accessible for guidelines on safe electrical work practices. Government electrical inspectors or GEIs play a key role in eliminating poor work practices carried out by electricians and contractors. Electrical inspections are normally done before temporary or permanent electrical connections are made to the utility's power supply.

Safety switches or disconnects are installed on electrical equipment to provide quick isolation of devices when necessary. These switches should be visible and easily accessible. Motor protection is required for all motors rated more than 1 A.

2.13 Test Your Knowledge

1. State three methods of providing mechanical protection for electrical cables.
2. State three types of protection which an electrical installation should offer.
3. Why is it important to earth all metal boxes in an installation?
4. Explain the words *shall* and *may*, which are found in electrical codes or reference material.
5. Why is it important to install light and outlets circuits in separate conduits?
6. If a distribution panel has ten circuits, how many neutral conductors are found in that panel?
7. Explain the dangers of sharing neutral conductors.
8. What are the conditions for using a spur to other appliances from a lighting circuit?
9. Explain the difference between lock rotor current and running current.
10. What are the safety conditions to be considered when making provisions for installation of a pad-mounted transformer?
11. What are the safety conditions to be considered when installing motors?
12. What is the purpose of the letter code found on the specification plate of a motor?
13. Explain the term *barriers*.
14. Explain the term *panel accessibility*.
15. Name three fixed appliances and three portable appliances.
16. A motor to be installed is rated at 15 hp, 3 phase, and has a letter code of K. What will be the starting current of that motor with an operating voltage of 208 V if the corresponding kVA for the letter code K is 8.99.
17. Using question 16, what will be the running current of this motor?

Biblography

I. National Fire Protection Association. 2008. National Electrical Code. National Fire Protection Association. Seventeenth ed. Pages 177-180

ii. Newnes Electrical Pocket Book. Torque/speed and current/speed curves for normal cage induction motor connected in star and delta. Twenty-third ed. Page 302

iii. The Institute of Electrical Engineers—IEE. 2008. Requirements for electrical installations; wiring regulations. The Institute of Electrical Engineers. Seventeenth ed. Pages 128-129

iv. Grainger. Greenlee PVC heating blanket 1994-2013. Retrieved on July 14, 2013, from http://www.grainger.com/Grainger/GREENLEE-PVC-Heating-Blanket-5C638

v. Lyncole XIT Grounding. Soil resistivity, testing four point Wenner method. Retrieved on July 11, 2011, from http://www.lyncole.com/uploads/Lyncole_Ground_Test_Methods.pdf

vi. IPEX. 2006. Scepter rigid PVC conduits & fittings, UL 90° Elbows. Retrieved on July 26, 2012, from http://electrolines.com/Images/Line%20Card/Catalogs/Rigid_PVC_Conduit_Fittings_Catalog.PDF

vii. TECHTIP 602001. Grounding and shielding considerations for thermocouples, strain gages, and low-level circuits. Retrieved on September 23, 2011, from http://www.mccdaq.com/pdfs/techtip/techtip_60201.pdf

CHAPTER 3

Circuit Protection

3.0 Introduction

Circuit protection is the action of isolating a circuit when unsafe operating voltages and currents are present. This involves installation of isolation mechanisms or devices such as fuses, circuit breakers, or relays with predetermined specifications in circuits. These mechanisms or devices monitor the voltage, current, and frequency and isolate when necessary to prevent electrocution and/or damage to property and equipment. Protective devices and monitoring equipment are designed to keep monitored values within safe operating and design limits.

Fuses, circuit breakers, and protective relays are a requirement for all electrical installation to provide the primary protection listed below:

1. Protect a circuit from sustained overcurrent
2. Protect a circuit from faults due to moisture
3. Protection against short circuit

All protective devices fall within one of the following groups:

1. Fuse Protection
2. Electromagnetic protection
3. Thermal protection—bimetallic only
4. Relay protection

3.1 Fuse Protection

Fuse protection has been used for many years for protecting cables, appliances, electronics, and many other electrical systems. Fuses are rated less than the ratings of circuit breakers and are used in both LV and HV systems. There are basically two types of fuses: cartridge and high rupturing capacity (HRC) cartridge.

3.1.1 Cartridge Fuses

Cartridge fuses are one of the most commonly used protective devices due to its ability to fit in small spaces. This type of fuse eliminates the use of robust enclosures used by other protective devices.

Cartridge fuses has been modernized to withstand greater short-circuit current and to provide greater flexibility in its use when compared with early designs. Typical cartridge fuses are shown in Figure 3.1.

(Adapted from Cooper Bussman, 2010)

Figure 3.1: *Typical cartridge fuses*

Cartridge fuses remain a significant protective device for many electrical installation and control systems. These fuses are used on LV system up to 600 V, range from 2-500 A, and have a breaking capacity up to 80 kA).

3.1.2 HRC Cartridge Fuses

HRC cartridge fuses are used in HV systems up to 11 kV and are available in ranges up to 350 A. HRC cartridge fuses are used to protect HV systems including transformers, capacitor banks, cables, and overhead power lines against short circuits. This type of fuse protects switch gears from thermal and electromagnetic effects of heavy short-circuit current, by limiting the peak current values and interrupting the currents in several milliseconds. A typical HRC fuse is shown in Figure 3.2.

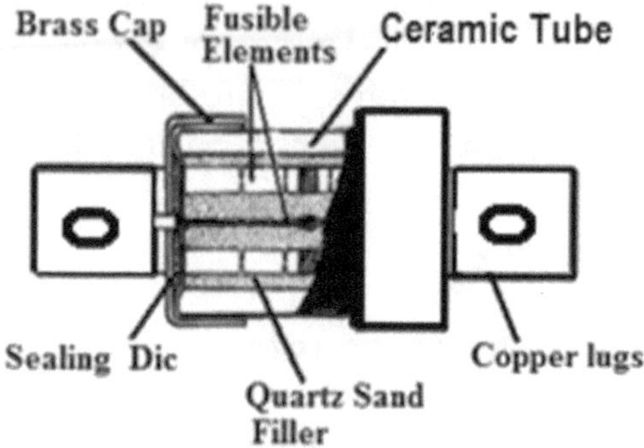

(Adapted from Versatile Protection with HRC Fuse Links, n.d.)
Figure 3.2: *HRC cartridge fuse*

3.1.3 Fuse Characteristics

The characteristic of a fuse dictates the speed at which the fuse raptures. Fuses are designed in three general categories; they are "extremely fast acting, fast acting, and slow acting." These characteristics technically are meaningfully referred to as the *I^2t value.* An I^2t value gives an appreciation of the current which passes through a fuse during a short circuit. This current is called "let-through current". Let-through current is created by a short circuit and can be extremely destructive. See Figure 3.3.

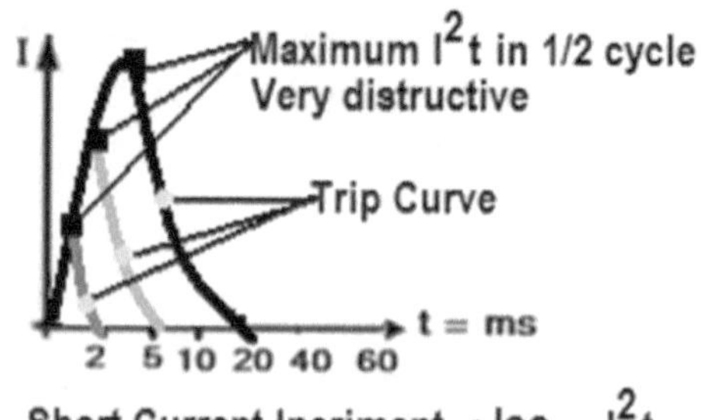

Figure 3.3: *Short-circuit incremental current*

3.2 Symmetrical and Asymmetrical Current

A symmetrical current is a uniform or normal balanced current among all phases in a circuit. If a circuit maintains the sinusoidal waveform supplied to it from a generator as shown in Figure 3.4, the circuit will remain symmetrical. Similarly, symmetrical faults current takes the path of the fundamental frequency. Symmetrical fault currents are normally created when loads are resistive. Conversely, asymmetrical (un-uniform) fault currents do not follow the path of the fundamental frequency and are created by most loads which are usually reactive.

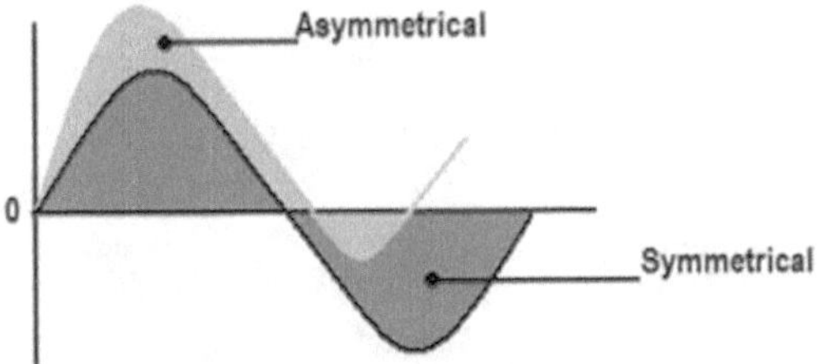

Figure 3.4: *Symmetrical and asymmetrical current waveforms*

Asymmetrical current is an imbalanced sinusoidal waveform as shown in Figure 3.4. This imbalance is caused by unbalanced load arrangement or short circuits or a fault which occurs unintentionally.

Protective devices operate in the normal state of symmetrical currents; once this principle is compromised by a fault, symmetrical current becomes asymmetrical current. This asymmetrical current pushes protective devices into trip mode. Asymmetrical currents created by a short circuit create a direct current (d.c.) component; this is the most destructive component of a short circuit. A short circuit current will include "Asymetrical Intrrupting Current (AIC), a "Root Mean Square" (r.m.s) component and a d.c. component. These components can be safely managed in a circuit.

As depicted in Figure 3.5, the d.c. component of an asymmetrical waveform is an exponential of a short-circuit current and is the most destructive component of any short circuit created. In addition, the d.c. component of a short circuit is similar to using a rectifier to converting an a.c. waveform to d.c. The only difference is, the d.c. component of a short circuit decays with time as shown in Figure 3.6 and a rectified a.c. waveform has a steady state. Protective devices such as circuit breakers are designed to withstand the d.c. component of short circuits.

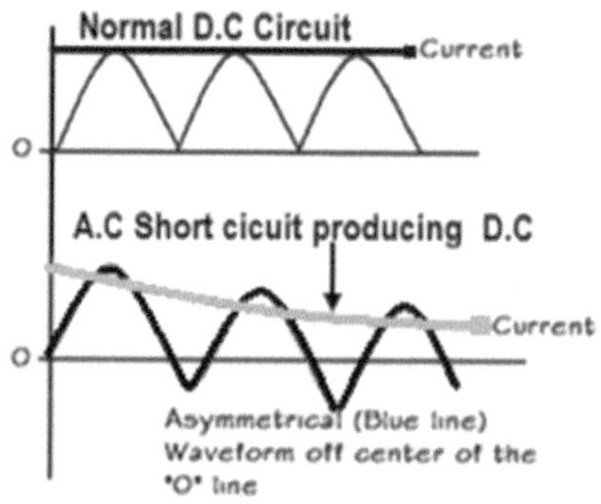

(Adapted from GE Application Information and Short-circuit Current Calculations)

Figure 3.5: *a.c. circuit producing d.c. current*

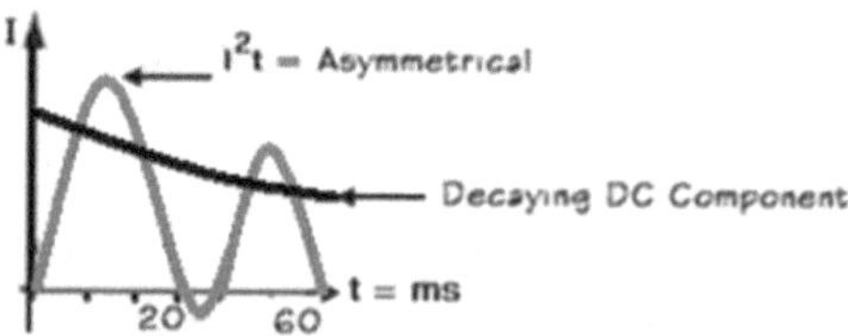

(Adapted from GE Application Information and Short-circuit Current Calculations)

Figure 3.6: *Comparing a decaying d.c. component to an a.c. asymmetrical component*

3.3 Circuit Breaker Protection

Circuit breakers are devices which are designed to isolate a circuit when a current which exceeds its predetermined value attempts to pass through it. A typical circuit breaker, as shown in Figure 3.7, is basically a sophisticated switch which automatically shuts off (trips) when the current passing through it reaches its trip value and can be reset to restore power in the isolated circuit.

(Adapted from Images for Square D single-phase circuit breakers, n.d.)

Figure 3.7: *Typical circuit breaker with overload and short-circuit specifications*

The rated current of a circuit breaker is not a representation of the magnitude of the short-circuit current available in an installation but rather a value determined by the rated current of the established circuit. For example, a #14 or a 1.5 mm^2 cable protected by a 15 A circuit breaker would be undersized if connected to a load of 20 A. However, the same cable would be sufficient to handle a fault current of 500 A during a short circuit for 0.8 Sec. Furthermore, if the fault is not cleared within the specified disconnection time, 500 A in the circuit would cause extensive damage to the cables and other circuits.

3.3.1 Elements of a Circuit Breaker (Bimetal)

A circuit breaker has a number of essential components as shown in Figure 3.8. Each of these components coupled to each other provides protection for an installation and individual circuits.

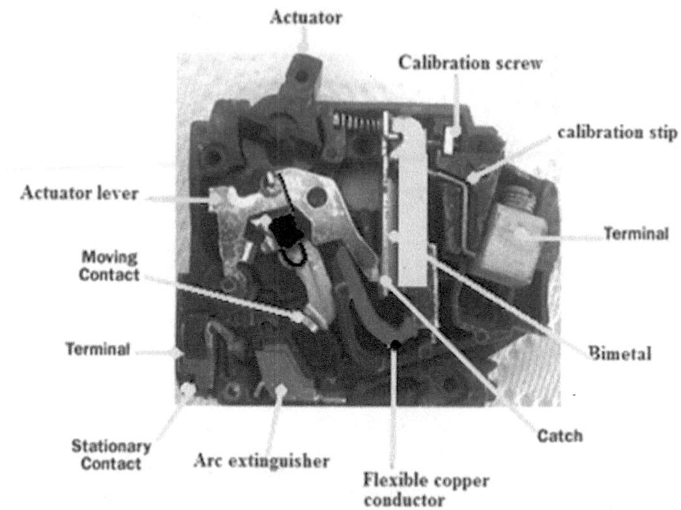

Figure 3.8: *Typical miniature thermal circuit breakers*

In this device, the functions of the various parts are as follows:

1. **Actuator lever**—Used to manually trip and reset the circuit breaker. It also indicates the status of the circuit breaker (on or off/tripped).
2. **Bimetallic strip**—Bends when heated by current passing through. When the current exceeds a certain value, the catch is released.
3. **Moving contacts**—Opens when there is trip by a fault or intentionally
4. **Terminals**—The circuit breaker has two terminals. One is for connecting the circuit conductors and the other for making contact (plugged or bolted) to the bus bar of the electrical panel.
5. **Catch—**Fitted on the bimetallic strip to set the actuator lever in its position.
6. **Calibration screw**—Precisely adjusts the trip current of the device after assembly.
7. **Arc extinguisher—**Suppresses the arcs which are produced when contacts are open on load or during a short circuit.

8. **Flexible copper conductor**—Transfers power between terminals.
9. **Calibration strip**—Controls the speed which the contacts open.
10. **Stationary contact—**Contact pushed or bolted to a bus bar.
11. **Moving contact**—Contact which opens and closes manually or automatically.

3.3.2 Electromagnetic Circuit Breakers

Electromagnetic protective devices provide protection for short circuit and overload conditions for all circuits in which they are installed. The tripping mechanisms in this device incorporate a bimetallic strip and an electromagnetic coil or solenoid. These tripping components are enclosed in a molded case which is able to withstand a high blast of arc and current, resulting from short circuit. Details of an electromagnetic circuit breaker are shown in Figure 3.9

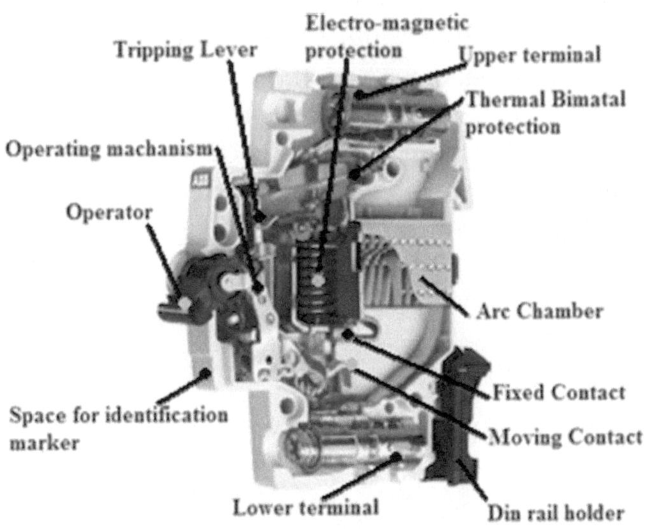

(Adapted from *Miniature Circuit Breakers Circuit Protection Guide*, n.d.)

Figure 3.9: *ABB current-limiting circuit breaker*

The ABB miniature circuit breaker operates on two principles: the overload current principle and the short-circuit principle.

The electromagnetic circuit in electromagnetic protective device is a simple solenoid which operates when a high fault current or short-circuit current passes through the solenoid. This type of device operates much faster than the miniature circuit breaker.

3.3.3 Thermal Circuit Breakers

Thermal protective devices are direct action devices which provide adequate protection for most LV installations. Thermal protective devices are selected based on one of the following criteria:

1. High initial starting or inductive current requirement
2. Normal load requirement

The word *thermal* signifies heat. Most miniature circuit breakers operate on the principle of metal expansion with heat. A bimetallic strip in miniature circuit breakers heats up and expands while high current passes through it. During the passage of fault current, the bimetal strips gradually expand and bend as a result of different expansion rates of two metals bonded together. The holding pin attached to the bimetal mechanism is displaced from its position, causing the circuit breaker to operate automatically. See Figures 3.10 and 3.11.

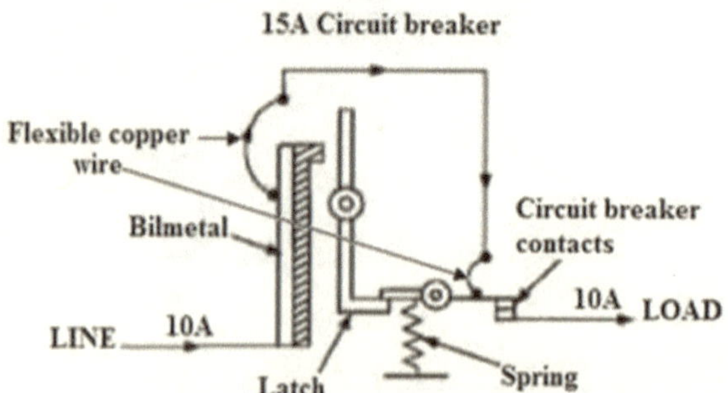

(Adapted from Schnieder Electric, Square D, October 2004)

Figure 3.10: *Circuit breaker operating under normal condition*

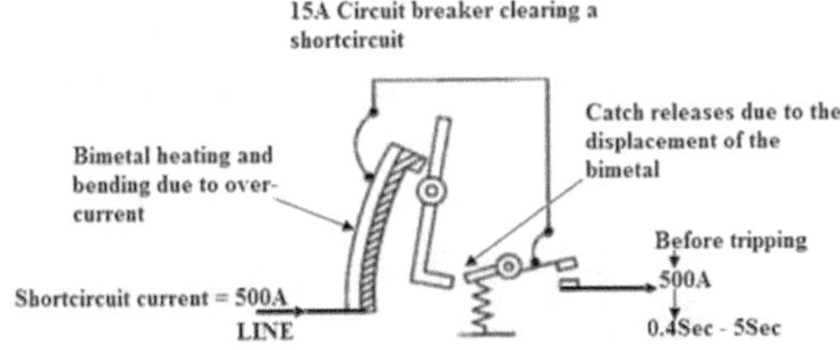

(Adapted from Schnieder Electric, Square D, October 2004)

Figure 3.11: *Circuit breaker experiencing a short-circuit current*

3.4 Circuit Breakers and High Inductive Current

A regular circuit breaker may not be adequate for installation where the high initial starting or inductive current is required; circuit breakers for this type of load relies on a high-sustained start-up current for a few milliseconds. Depending on the lock rotor ampere (LRA) or the start-up current of the device, a standard circuit breaker will trip while starting. The circuit breaker for these installations should be fitted with adjustable bimetal time delay trip.

In Figure 3.12, the starting current of a motor rises from zero to 40 A and then fall to 18 A. Consequentially, if a circuit breaker rated at 20 A was to be installed without a time delay, this circuit would trip while starting the motor.

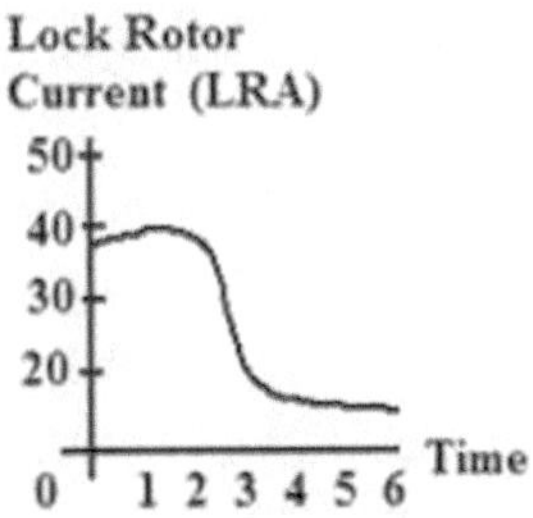

Figure 3.12: *Typical LRA characteristics for an induction motor*

Time delays trip fitted on circuit breakers is designed to hold high start-up current of motors for a few seconds, allowing the current to fall to normal running current. However, circuit breakers will trip if the starting current of a motor is held longer than the time set on the time delay.

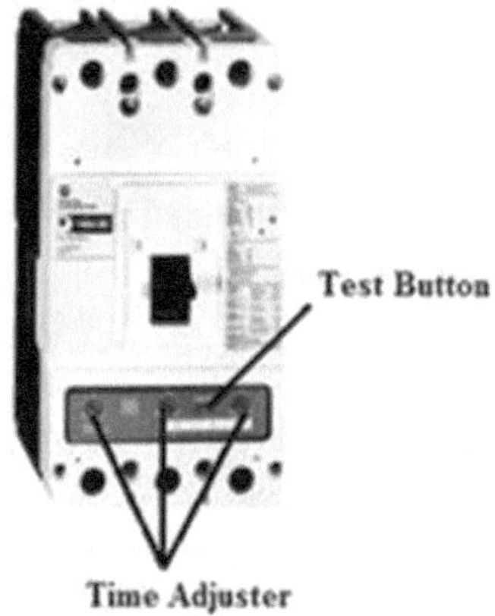

(Adapted from Rockwell Automation, October 2007)

***Figure 3.13**: Three-phase circuit breaker fitted with time delay trip*

The time adjusters seen in Figure 3.13 are used to adjust the sensitivity of each phase. This means that the circuit breaker rated at 400 A as shown in the figure, trips at 3 seconds after experiencing the fault current of 400 A.

If time delay circuit breakers are used for protecting installations where motors and other high inductive circuits are not a part of the installation, time adjusters should be set at the minimum or zero. This will allow the circuit breaker to operate as a normal circuit breaker. However, it is not recommended that time delay circuit breakers be installed where they are not required since they are significantly more expensive than regular circuit breakers.

Later in this text, further information on selecting circuit breakers will be introduced and discussed.

3.5 Why Protect Cables?

Electrical cables, whether main supply cables or sub-panel supply cables, are affected by the short-circuit current which are developed on sub-circuits such as power outlets or motor circuits. The effect a short circuit has on cables may not be significant to be noticed if the correct cable sizes are installed. However, an awareness that changes place in these cables is extremely crucial. Since both temperature and current changes occur in main cables, appropriate protective devices must be selected with minimum over tolerance to reduce the possibility of cable damage as depicted in Figure 3.14.

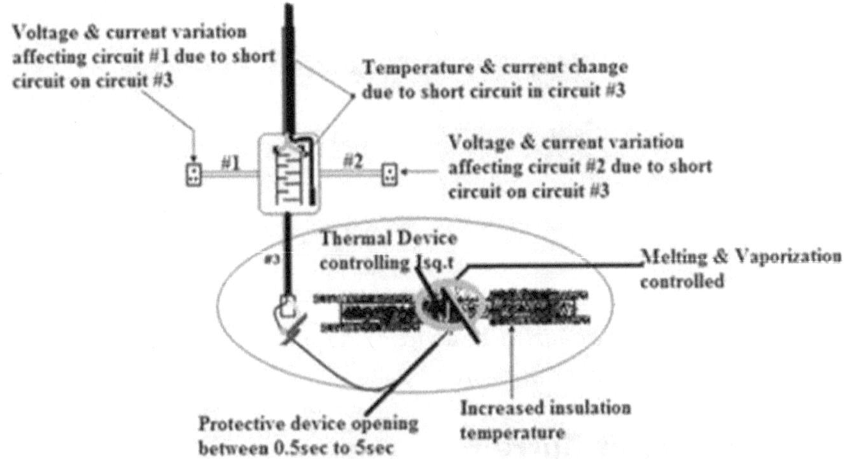

Figure 3.14: *The effect of localized short circuit in conductors and cable insulation*

The main cable size of an installation should not be less than four times the sum of the largest current using equipment or four times the sum of the total load to be used at the same time. The total calculated current installations should be divided by 1.414 for single-phase installation and 1.732 for three-phase installations. The computed figure will be the total balanced current carried by each phase conductor. Computation of the size of main and circuit cables is essential since this will avoid under-sizing of cables and hence reduce the effect of localized heating of conductors during a short circuit.

3.5.1 Localized Heating Effect of Conductors

Localized heating effect is the destructive heat which a circuit develops during a short circuit, for example, phase-to-ground or phase-to-phase faults. Furthermore, the heat generated by a short circuit affects not only the short-circuited circuit itself but other circuits which are bunched with the shorted cable as shown in Figure 3.15. This condition will result in total ruin of all other cables bunched together or inside the same raceway or conduit.

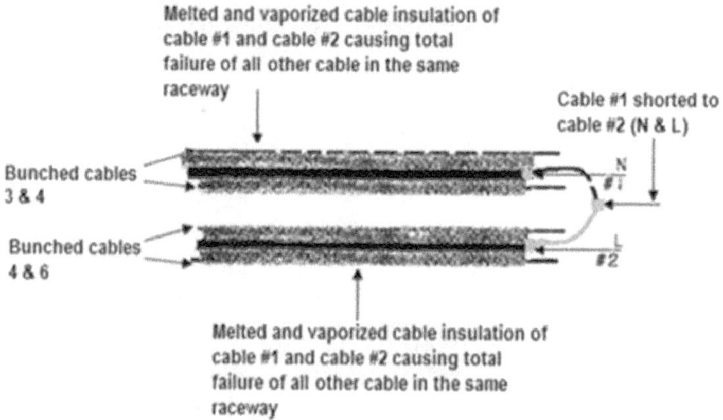

Figure 3.15: *Localized heat from two shorted cables (live and neutral)*

3.6 Short-circuit Current

Short-circuit currents on electrical installations are produced and delivered by distribution transformers supplying the installations. The value of a short-circuit current is dependent on the power rating of the transformer supplying the installation and the time the circuit breaker takes to clear a fault or a short. As a result, the cables become overburden during a short circuit, whether it is a phase-to-earth or phase neutral fault. This will cause loose termination or joints to be blown off during a short circuit.

An analysis of a short circuit created between two cables, *phase (line 1) and neutral*, is shown in Table 3.1. This table provides a practical demonstration of the behavior, the effect, and the importance of clipping or isolating short-circuit current quickly. This table also shows the significant effect a short-circuit created on one phase could have on the other phases of the installation, for example, *line 2* and *line 3*.

In Table 3.1, a fluke analyzer was used to analyze the behavior of the current and voltage on *line 1* and *line 2* during the short circuit.

Table 3.1: Short-circuit study

Test Instrument—Fluke 434 Power Quality Analyzer	
Line 2—Experiencing unbalance current due to short on L1	
Circuit current before short circuit	21 A
Short-circuit current	35 A
Voltage before short circuit	123 V
Disconnection/clip time of circuit breaker	2.5 ms
Neutral circuit current before short circuit	0 A
Line 1—Shorted to neutral	
Short-circuit clipped current I^2t or asymmetrical	270 A
Available short-circuit current at receptacle	500 A
Line voltage before short	124 V
Short-circuit clipped voltage	1.1 V
Neutral current	18 A
Temperature of shorted cable insulation	Warm
Type of conductor	Stranded
Conductor size	#12 or 2.5 mm
Type of conductor	XLPE
Condition of shorted conductor	Cable strands fused together

This study shows that within a few milliseconds, a short circuit could cause the asymmetrical current of a 20 A circuit to spiral to hundreds or thousands of amperes. The study also proves that a total imbalance of a distribution system could cause sensitive equipment to malfunction and become damaged by the absorption of high current generated in the circuit. Careful analysis of Table 3.1 reveals that clipping or stopping of fault within the first 10 ms for circuits with sensitive equipment is critical to the protection of life, property, and equipment.

3.6.1 Short-circuit Energy

According to ABB Inc, "During a short circuit, both magnetic forces and thermal energy combine to damage electrical devices on the network. The level of thermal energy and magnetic forces is directly proportional to the square of the current."

Controlling the *thermal energy* produced during a short circuit, synchronizes with controlling the I^2t or the asymmetrical current when a short is created. Therefore, to control the *thermal energy* produced during a short circuit, it is important to be aware of the amount of heat dissipated in cable insulations when short circuits are created. The heat dissipated in cable insulations during a short circuit can be compared to the heat generated by a 100-watt incandescent bulb. It can therefore be assumed that during a short circuit, cables act as electrical heating elements and become self destructive. The energy generated is measured in Joules/second. For a 100W bulb, the energy generated is 100 Joules/sec.

3.6.2 Thermal Energies of Copper

During a short circuit, copper conductors melt quickly and vaporize if I^2t is not controlled quickly. Table 3.2 provides an appreciation of the thermal characteristics of copper.

Table 3.2: Thermal specifications of copper

Properties	Values
Specific heat capacity	0.385 J/s
Heat of fusion	204.8 J/s
Energy required to melt	630 J/s
Melting point	1083° C
Boiling point	2580° C
Heat of vaporization	5234 J/s

3.7 Structural Damage of Conductors

Figures 3.16 and 3.17 provide examples of damages done to copper conductors and by extension the entire circuit, when a short circuit occurs and the thermal specifications of copper are exceeded.

Figure 3.16 and Table 3.2 show that the power released during a short circuit is equivalent to the square of the current and can result in "fault power, equivalent to thousands of horsepower with its arc temperatures as high as 6000° C." This temperature is significant as it shows how quickly a copper conductor can vaporize. As shown in Table 3.2, copper vaporizes at *5234 J/g*. It is for this reason that clearing a fault or a short circuit within the quickest possible time is critical.

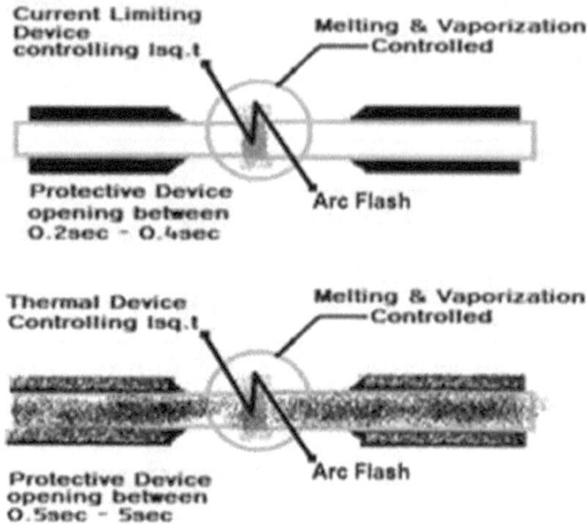

Figure 3.16: *Incremental damage versus time*

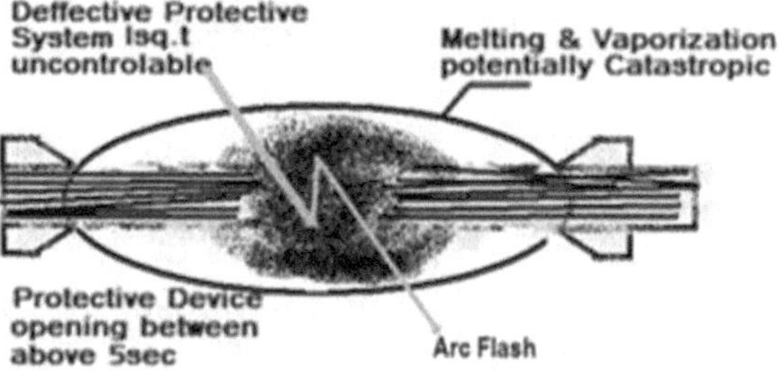

Figure 3.17: *Continued incremental damage over time*

3.8 Selecting Circuit Breakers Based on AIC or r.m.s Values

Fitting the correct circuit breaker on an installation extends the service life of the circuit on the installation. However, AIC selection of circuit breakers should be based on the design drawings with values already calculated or on transformer specification. In addition, an installation with its available short-circuit current up to 6 kA should be fitted with circuit breakers having asymmetrical current ratings (AIC/r.m.s) not less than 10 kA.

In an installation designed to be supplied by a transformer with an available short-circuit current of 16 kA, the AIC rating of the main circuit breaker should not be less than 16 kA if the transformer is installed within 50 feet of the installation. Consequently, if the kA rating of the main circuit breaker installed in this installation is far less than the AIC rating of the transformer, for example 10 kA, the circuit breaker would burn out quickly after a number of short circuits.

Circuit breakers have two functions: to operate on overload conditions and to operate in short-circuit conditions. See Table 3.3.

Table 3.3 shows the specification of spectra circuit breakers with the following abbreviations"

1. *ICU* is the ultimate or final short-circuit breaking capacity
2. *ICS* is the service short-circuit breaking capacity

Where *IC* represents the interrupting current, *U* represents ultimate or final current, and *S* represents service current.

Table 3.3: Specification of Spectra Circuit Breakers

Type	Ampere Rating	Poles	220 - 240 Vac		380 - 415 Vac		500 Vac		690 Vac	
			Icu	Ics	Icu	Ics	Icu	Ics	Icu	Ics
SED	15 - 32	2	18	9	10	5	-	-	-	-
		3					4	4		
SEH		2	65	33	15	10	-	-	-	-
		3					6	6		
SEL		2	100	50	20	15	-	-	-	-
		3					8	8	3	3
SEP		2	200	100	20	20	-	-	-	-
		3					10	10	5	5
SED	40 - 160	2	18	9	14	7	-	-	-	-
		3					14	7		
SEH		2	65	33	35	17	-	-	-	-
		3					25	12		
SEL		2	100	50	65	33	-	-	-	-
		3					40	20	5	5
SEP		2	200	100	100	50	-	-	-	-
		3					50	25	10	5
SFH	70 - 250	2	65	33	35	17	-	-	-	-
		3					25	12		
SFL		2	100	50	65	33	-	-	-	-
		3					40	20	14	7
SFP		2	200	100	100	50	-	-	-	-
		3					65	33	18	9
SGH	125 - 600	2	65	33	25	13	-	-	-	-
		3					18	9		
SGL		2	100	50	65	33	-	-	-	-
		3					35	18	14	7
SGP		2	200	100	100	50	-	-	-	-
		3					50	25	18	9
SKH	300 - 1250	2	65	16	50	13	25	13	-	-
		3								
SKL		2	100	25	65	16①	42	21①	14	14
		3								
SKP		2	140	35	85①	25	50	25	18	18
		3								

3.9 Conclusion

Protective devices are designed to isolate circuits when a current exceeds its predetermined value. These predetermined values are referred to as the normal operating current of a circuit or installation.

Circuit breakers operate on two principles: overload current and short-circuit current. The overload current value labeled on circuit breakers is sometimes misunderstood to be the representation of the magnitude of the available short-circuit current in an installation. Overload currents are represented by ampere (A) values and short-circuit currents are represented by kA.

The increase of the thermal energy developed on cable insulations is also controlled by fuses and circuit breakers. Therefore, since the temperature on a circuit breaker is in proportion to time, it becomes critical not to overload circuits or load circuits close to its full load capacity.

3.10 Test Your Knowledge

1. List three primary functions of protective devices.
2. List three groups of protective device.
3. Draw and label an HRC fuse.
4. Explain the significant difference between a 50 A 22 kA circuit breaker and a 50 A 10 kA circuit breaker.
5. A 50 A circuit breaker is incorrectly placed on a 20 A circuit. If there were to be a short circuit, explain what effect if any would this have on the circuit?
6. A 70 A circuit breaker is incorrectly placed on a 15 A circuit within the same panel. If there was to be an overload on this circuit, explain what effect this would have on the appliance or equipment which is experiencing this overload?
7. What is the difference between the let-through current of a thermal circuit breaker and a current-limiting circuit breaker?
8. List three main components of the current-limiting circuit breaker.
9. When selecting thermal circuit breakers, list three factors or conditions to be taken into consideration.
10. List five components of a miniature circuit breaker.
11. What is the purpose of the thermal adjusters and the test button found on some circuit breakers?
12. Describe the protection which can be provided by relays.
13. Name three characteristics of a relay.
14. Why is it important that the AIC value of a circuit breaker is greater than the available short circuit current on the installation?
15. Why is it important to install supplementary earth electrodes?
16. Why is it important to avoid loading a transformer to full capacity?
17. Define the terms *r.m.s.* and *AIC value*.
18. Why are r.m.s. or AIC values written on circuit breakers?

Bibliography

i. ABB. Medium Voltage HRC Fuses. Catalogue sheet B30/06.03e. Edition, 08, 2001. Retrieved in October 2011, from http://www05.abb.com/global/scot/scot235.nsf/veritydisplay/39ae25c019313b2ec1256c28003d032a/$file/bwmwang.pdf

ii. Appendix B—Asymmetrical Current. n.d. Retrieved in November 2010, from http://www.hubbellpowersystems.com/literature/encyclopedia-grounding/pdfs/07-0801-appendix-b.pdf

iii. Copper Bussmann. n.d. Low Voltage Branch Circuit Rated Fuses. Retrieved in November 2013, from http://www1.cooperbussmann.com/pdf/25f32cf8-913b-4934-8a5f-3b404d08e890.pdf

iv. GE Electrical Distribution & Controls. Application information, Short-circuit Current Calculations. n.d. Retrieved in October 2011, from http://www.geindustrial.com/publibrary/checkout/GET-3550F?TNR=White%20Papers|GET-3550F|generic

v. GE Consumers & Industrial Electrical Distribution. 04, 2008. Spectra RMS™ Molded Case Circuit Breakers. Retrieved in September 2011, from http://www.geindustrial.com/publibrary/checkout/GET-7002D?TNR=Application%20and%20Technical|GET-7002D|generic

vi. Google. Images for Square D Single Phase Circuit Breakers. n.d. Retrieved in October 2011, from https://www.google.com/search?q=Images+for+square+D+single+phase+circuit+breaker&tbm=isch&tbo=u&source=univ&sa=X&ei=8PDrUc3DN4Xc8wTFklCoCw&ved=0CCwQsAQ&biw=1366&bih=667

vii. Mike Holt Enterprise. 2003. National Electrical Code Internet Connection. Available Short-Circuit Current. Retrieved on November 2010, from http://www.mikeholt.com/mojonewsarchive/EES-HTML/HTML/ElectricalCircuitBreakers~20030621.htm

viii. Miniature Circuit Breakers. *Circuit Protection Guide*. n.d. Retrieved on August 2011, from http://www05.abb.com/global/scot/scot209.nsf/veritydisplay/ef5f3a6e43c99c288525758500534fe6/$file/1sxu400142m0201.

ix. Schnieder Electric. *HV Training Manual*, June 2003 [Percentage Impedance]

x. Schnieder Electric. October 2004. Square D, Thermal-magnetic/Magnetic Only Moulded Case Circuit Breakers. Retrieved on June 10, 2013, from http://stevenengineering.com/tech_support/PDFs/45CBTHER.pdf

xi. Williams, P. T. Versatile Protection with HRC Fuse Links. n.d. Retrieved on July 4, 2013, from: http://www.fuseco.com.au/files/General_Fuseology_articles/Versatile_protection_with_HRC_fuse_links.pdf

CHAPTER 4

Fault Current and Transformers

4.0 Introduction

A transformer is a unique electrical device in electrical power engineering and is the *recipient and producer* of all fault currents. This device is critical in the transformation and the delivery of electrical power from very small wattage to very large megawatts. Voltages of transformers vary in accordance with its design, use, purpose, and power requirements. Table 4.1 shows a number of transformers which are found in industries, electronics, and household appliances. They are easily identified because of their size and functions.

Table 4.1: Types of Transformers

Dry Type	Wet Type (Oil)
Pad-mounted	Pole and pad-mounted
-	Power
Distribution	Distribution
3-phase	3-phase
Single phase	Single phase
1:1 ratio/isolation	-
Delta/star	Delta/star
Star/delta	Star/delta
Auto transformers	-
Signal	-

4.1 Transformers Windings

Transformers are designed with two windings: primary and secondary. These windings are wound on hundreds of laminated (coated for insulation and vibration reduction) pieces of metal, pressed and welded or bolted together to form a core which is in some cases called a former. A transformer core/former is shown in Figure 4.1. Most transformers operate on the principle of mutual inductance. As shown in Figure 4.2, mutual inductance occurs when a magnetic field produced in the primary winding cuts across the secondary coil and creates a potential in the secondary windings.

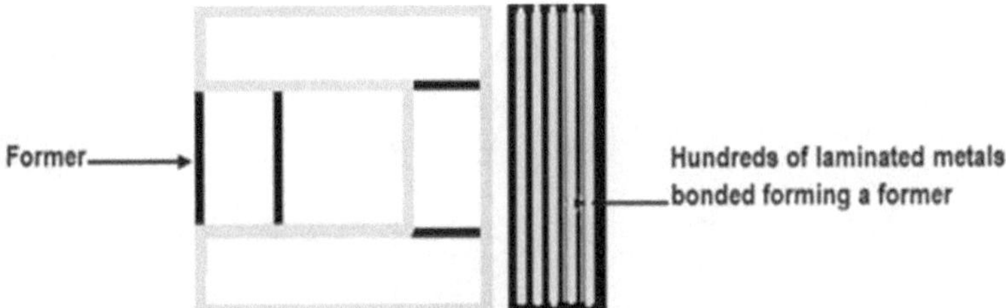

Figure 4.1: *Typical transformer core/former*

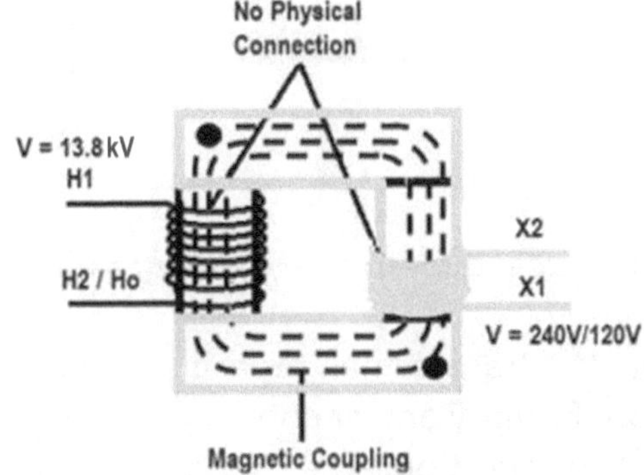

Figure 4.2: *Mutual inductance in transformers*

A typical example of a single-phase transformer core is shown in Figure 4.2. It is essential that electricians fully understand the construction of transformers and their winding configuration to fully understand the development and behavior of fault current in this device. This also holds true for three-phase transformers which is widely used in electrical power systems.

A typical pad-mounted three-phase transformer is shown in Figure 4.3. This device has three pairs of windings (6 coils). The voltage of each phase of a three-phase transformer is electrically displaced by 120^0 and is represented by two pairs of windings wounded on the same yoke of the former or iron core. One coil represents the high side or the primary side (H) and the other represents the low side or the secondary (X). The same is true for pole-mounted single-phase transformers which are configured to give the same outcome.

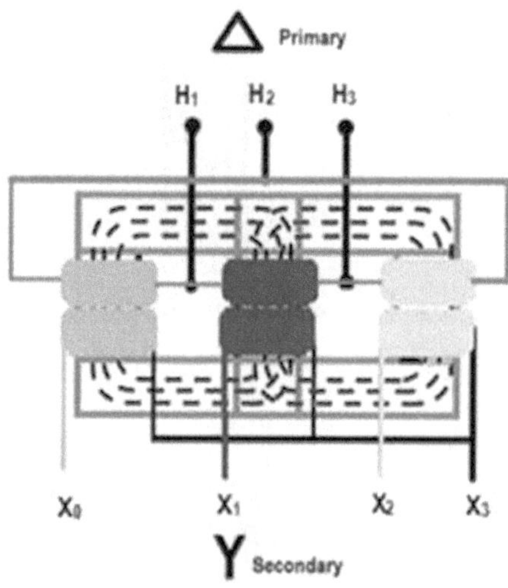

Figure 4.3: *Typical three-phase delta/star transformer*

The construction of a transformer, regardless of its size, is similar in core construction. Primary and secondary windings are connected in the desired configuration. The behavior of the fault current remains the same; however, fault power levels of all transformers are different. It is for this reason that understanding the principles of fault current is very critical in the field of electricity.

4.1.1 Transformers Winding Configurations

There are four configurations for three-phase transformers. They are as follows and as shown in Figure 4.4:

1. Delta/delta
   (has no neutral point; line voltage equals phase voltage $E_L = Ep$ and fault current is contained within the winding of the transformer.)
2. Star/star
   (has a neutral point for both primary and secondary windings; phase voltage equals line voltage $Ep = E_L / \sqrt{3}$; fault current flows through the system including the neutral and is not contained within the winding of the transformer.)
3. Delta/star
   (neutral is available only in the secondary winding.)
4. Star/delta
   (neutral is available only in the primary winding.)

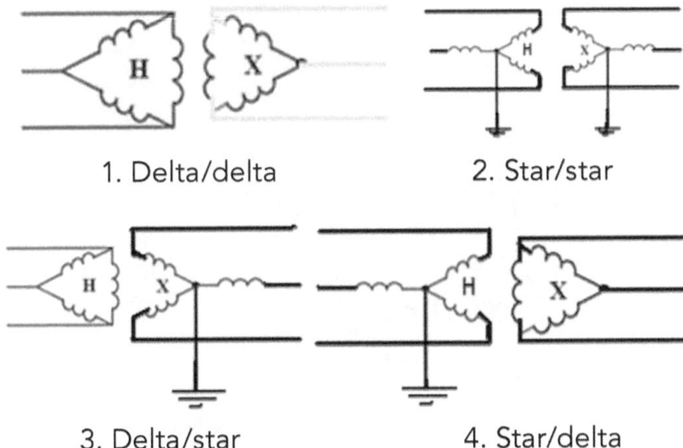

Figure 4.4: *Transformer configurations*

4.2 Prospective Fault Current

"Prospective fault current (PFC) is the maximum fault current or over-current which is developed at the point of contact when a short circuit occurs between a phase and a neutral conductor."

The magnitude of this current relies on the impedance and reactance of the affected cable and the short-circuit current available in the transformer.

4.2.1 Cable Reactance

All electrical conductors due to their material composition, have a characteristic resistance which opposes the flow of current. The resistance of a cable will increase as a result of variation in its length and cross sectional area (CSA) since the resistance is relative to these parameters. When the conductor is subjected to an alternating current (a.c.), the resistive characteristic is termed *reactance* and is represented by the abbreviation X_L.

Power cables of different phase (*L*1, *L*2, *L*3, *N*, or *G*) or (*L*1, *L*2, *N*, or *G)* in the same conduit will cancel expansion and circulation of the magnetic produce inductive reactance (X_L) in each phase. The cancelation of flux will reduce the effect of reactance on power cables. Figures 4.5 and 4.6 show the behavior of the magnetic forces between two parallel cables.

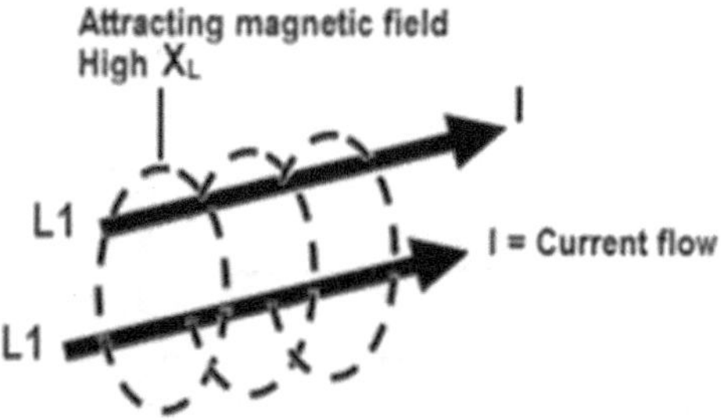

Figure 4.5: *Cables attracting due to high X_L*

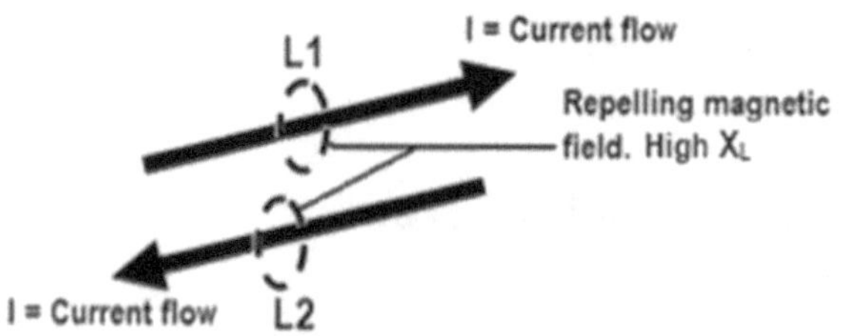

Figure 4.6: *Cables repelling each other due to high X_L*

4.3 Transformer Percentage Impedance

The percentage impedance (%Z) of a transformer is the percentage voltage required to raise the secondary current from zero to the designed secondary current when the secondary windings are short-circuited.

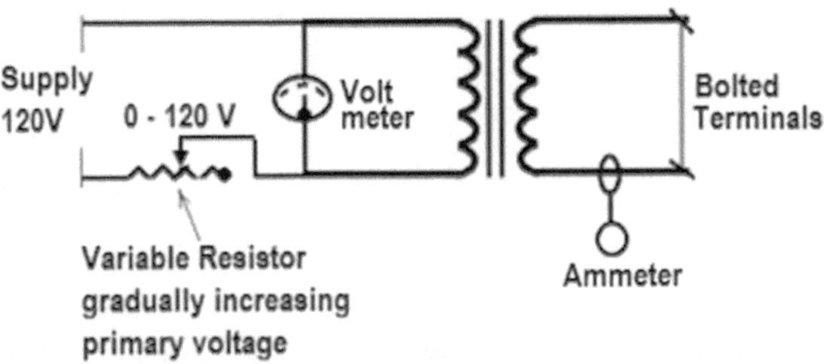

Figure 4.7: *A typical non-destructive test for determining the %Z of a transformer*

To determine the characteristics of the transformer, a short-circuit test as depicted in Figure 4.7 is done by manufacturers using specialized precision equipment. The test will reveal that transformers with low-impedance values produce more destructive power than those with high-impedance values. Furthermore, it is essential to note that the further away an electrical installation is from the supply transformer, the lower the destructive power will be on the installation or at the point where the short circuit is created because of volt drops in the connecting wires. The closer the transformer is to the installation, the higher will be the short-circuit current at that installation. See Figure 4.8 and Example 4.1.

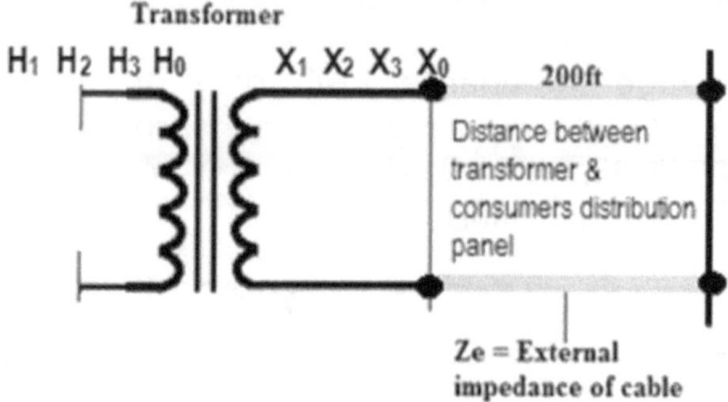

Figure 4.8: *Implication of distance on short-circuit current*

Example 4.1:

An installation is supplied by a 225 kVA 208 V three-phase transformer with 5.5% impedance. What is the value of the short-circuit current produced by this transformer if the transformer efficiency is 99%?

The impedance of utility transformers is rated in percentage (%) impedance.

Available Short-circuit current (*Sci*) at the transformer = I / %*Z*

$$\% \text{ Impedance } (\%Z) = \frac{5.5}{100} = 0.055$$

0.055—the percentage of the supply voltage required to raise the primary current to the rated current of the transformer when the terminals are short-circuited.

$$I_{line} = \frac{225\text{kVA}}{208 \text{ x}0.99 \text{ x}\sqrt{3}} = 631 \text{ A}$$

Full load current @ 100% efficiency

$$I = 631 \text{ A}$$

$$S_{ci} = \frac{\text{Tran. full load current}}{\%\text{Z}}$$

$$= \frac{631}{0.055}$$

Available at transformer S_{ci} is 11,473 A

Electrical cables will allow thousands of amperes to flow during a short-circuit; however, the increase in the short-circuit current flowing in an installation is controlled by the impedance of the cable, the power of the utility transformer, and the duration of short-circuit currents.

4.3.1 The Effect of Cable Reactance on Short-circuit Current

The total short-circuit current on an installation is dependent on a multiplier which known as the X/R ratio between the transformer, the MDPs, and the sub-circuits including generators and motors.

> "In some short circuit studies, the X/R ratio is ignored when comparing the short circuit rating of the equipment to the available fault current at the equipment."

It is important to note that LV gear (circuit breakers, switch gear, and other equipment) is tested based on X/R ratios. "X/R ratio is important because it determines the peak asymmetrical fault current. The asymmetrical fault current can be much larger than the symmetrical fault current."

Example 4.2 below clarifies short-circuit current including X/R ratio. Short-circuit current is classified in three categories in this section; they are three-phase, single-phase line-to-line, and single-phase line-to-neutral or ground.

$$\textbf{3 phase faults} = \frac{1.732 \text{ x L x } I_{3\emptyset}}{\text{C x n x } E_{L\text{-}L}}$$

$\textbf{Single phase Line} - \textbf{line}(L - L)\textbf{faults}$ See Tables 4.2 and 4.3

$$= \frac{2 \text{ x L x } I_{L\text{-}L}}{\text{C x n x } E_{L\text{-}L}}$$

$\textbf{Single phase Line} - \textbf{Neutral}(L - N)\textbf{faults}$. See Tables 4.2 and 4.3

$$= \frac{\text{L x } E_{L\text{-}N}}{\text{C x n x } E_{L\text{-}N}}$$

Where

L = Length of conductor (ft.)
C = Constant or "k" values for conductors See Table 4.2
N = Number of conductors per phase (adjust *C* values for parallel runs)
I = Available short-circuit current in amperes in the transformer to the end of circuit cable
E = Voltage

Example 4.2

The service for an installation is 500 MCM × 2 per phase as shown in Figure 4.9. The installation is located 200 feet from the supply transformer and is designed to supply load details as shown in Figure 4.9. The supply transformer is rated at 225 kVA 208 V three-phase and %Z = 5.5%. Taking into consideration X/R ratio between the transformer and each point of the installation, what is the value of the short-circuit current at each point of the installation?

Note: Short-circuit current is directly proportional to the cable and bus bar sizes. See Tables 4.2 and 4.3 for cable and bus bar constant.

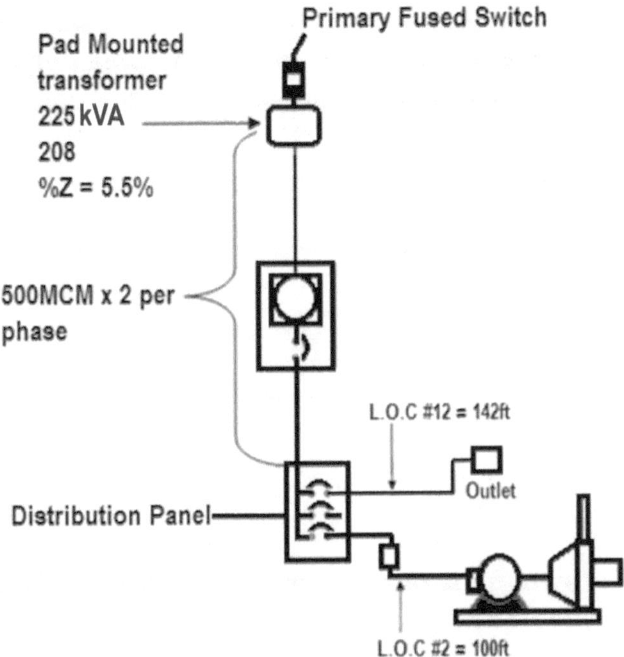

Figure 4.9: *Layout for example 4.2*

Step 1: *Calculate full load current*

$$\frac{225 \text{ x } 1000}{208 \text{ x } 1.732} = 625\text{A}$$

Step 2: *Calculate the multiplier using %Z of the transformer*

$$\frac{100}{5.5} = 18.18\ \Omega$$

Step 3: *Calculate transformer short-circuit let-through current*

Full load current x Multiplier

I_{SC}.= 625 x 18.18 = 11,364 A

I_{SC} motor = 4 x 625* = 2500 A

Total I_{SC} sym r.m.s. = 11,364 + 2500 = 13,864 A

Step 4: *Calculate the f factor, taking into consideration cable specification (length, number of cable per phase, and cable constant).* See Tables 4.2 & 4.3.

$$f = \frac{1.732 \text{ x } 200 \text{ x } 13{,}864}{18{,}177 \text{ x } 2 \text{ x } 208}$$

=0.064

Step 5: *Calculate multiplier M using f factor*

$$M = \frac{1}{1+f}$$

$$= \frac{1}{1+0.36} = 0.74$$

Step 6: *ISC sym r.m.s. due to cable X/R ratio = 13,864 x 0.74 = 10,259.36 A*

*I*SC motor contribution = 4 x 625* = 2,500 A

I total SC sym RMS = 10,259.36 + 2500 = 12,759.36 A

Repeat steps 4 to 6 for each subsection or panel of the installation

Step 4: *Calculate the f factor, taking into consideration cable specification (length, number of cable per phase, and cable constant).* See Tables 4.2 and 4.3.

$$f = \frac{1.732 \text{ x } 142 \text{ x } 13{,}864}{617 \text{ x } 208} = 26.5$$

Step 5: Calculate multiplier *M* using *f* factor

$$M = \frac{1}{1+f}$$

$$= \frac{1}{1+26.5} = 0.038$$

Step 6: I_{sc} sym r.m.s. = 13,864 x 0.038 = ***526.81 A***

Figure 4.10 shows a distribution system which incorporates Examples 4.1, 4.2, and 4.3. The figure also shows that cables with high impedance (small cables) will conduct a small amount of the short-circuit current produced by the distribution transformer, and that cables with low impedance (large cables) will conduct a huge amount of the short-circuit current produced by the distribution transformer. In addition, a circuit will experience maximum short circuit if the circuit breaker does not trip or clear the fault within 0.4 seconds.

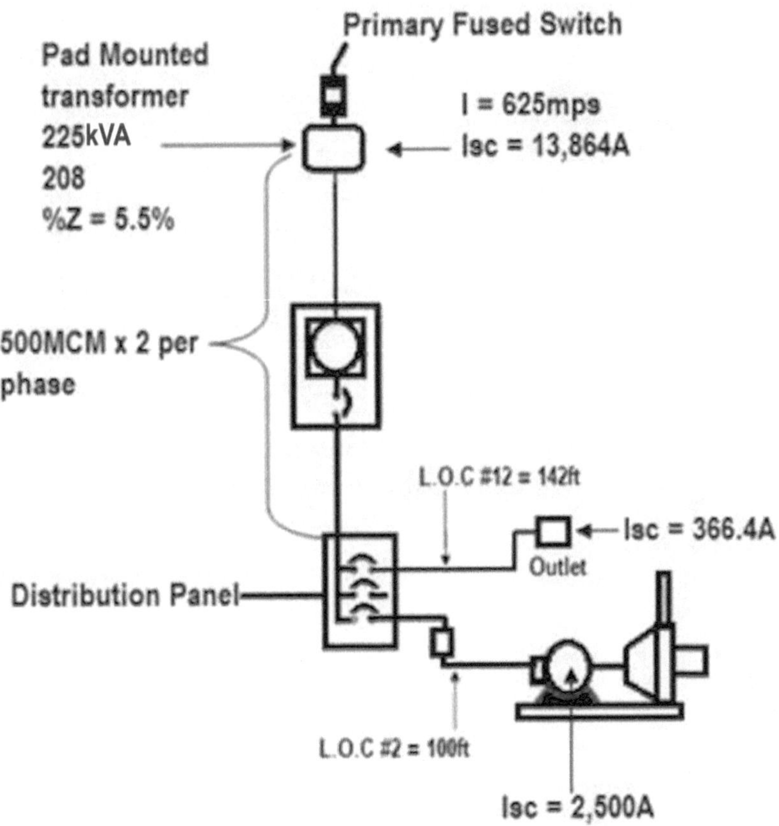

Figure 4.10: *Short circuit result using X/R ratio short-circuit current*

Table 4.2: Constant for single conductors in conduit

Copper AWG or kcmil	Three Single Conductors Conduit Steel			Nonmagnetic		
	600V	5kV	15kV	600V	5kV	15kV
14	389	-	-	389	-	-
12	617	-	-	617	-	-
10	981	-	-	982	-	-
8	1557	1551	-	1559	1555	-
6	2425	2406	2389	2430	2418	2407
4	3806	3751	3696	3826	3789	3753
3	4774	4674	4577	4811	4745	4679
2	5907	5736	5574	6044	5926	5809
1	7293	7029	6759	7493	7307	7109
1/0	8925	8544	7973	9317	9034	8590
2/0	10755	10062	9390	11424	10878	10319
3/0	12844	11804	11022	13923	13048	12360
4/0	15082	13606	12543	16673	15351	14347
250	16483	14925	13644	18594	17121	15866
300	18177	16293	14769	20868	18975	17409
350	19704	17385	15678	22737	20526	18672
400	20566	18235	16366	24297	21786	19731
500	22185	19172	17492	26706	23277	21330
600	22965	20567	17962	28033	25204	22097
750	24137	21387	18889	29735	26453	23408
1,000	25278	22539	19923	31491	28083	24887

Table 4.3: Constant for three-conductor in conduit

Copper						
AWG or kcmil	Three-Conductor Cable Conduit					
	Steel			Nonmagnetic		
	600V	5kV	15kV	600V	5kV	15kV
14	389	-	-	389	-	-
12	617	-	-	617	-	-
10	982	-	-	982	-	-
8	1559	1557	-	1560	1558	-
6	2431	2425	2415	2433	2428	2421
4	3830	3812	3779	3838	3823	3798
3	4820	4785	4726	4833	4803	4762
2	5989	5930	5828	6087	6023	5958
1	7454	7365	7189	7579	7507	7364
1/0	9210	9086	8708	9473	9373	9053
2/0	11245	11045	10500	11703	11529	11053
3/0	13656	13333	12613	14410	14119	13462
4/0	16392	15890	14813	17483	17020	16013
250	18311	17851	16466	19779	19352	18001
300	20617	20052	18319	22525	21938	20163
350	22646	21914	19821	24904	24126	21982
400	24253	23372	21042	26916	26044	23518
500	26980	25449	23126	30096	28712	25916
600	28752	27975	24897	32154	31258	27766
750	31051	30024	26933	34605	33315	29735
1,000	33864	32689	29320	37197	35749	31959

Courtesy of Cooper Industries, Copyright © 2011 Cooper Industries

4.4 Conclusion

Material composition of electrical conductors has a resistance which opposes the flow of current during a short circuit. This opposition is called reactance and is represented by the abbreviation X_L.

The voltage to current ratio of a transformer refers to the %Z of a transformer. This voltage to current ratio determines the maximum short-circuit current found in an installation.

The magnitude of short-circuit current can be catastrophic; therefore, limiting the current created by a short should become critical during the design stage of an installation. It is imperative that the %Z of a transformer be kept to the lowest possible value.

4.5 Test Your Knowledge

1. What is PFC?
2. List two methods of identifying the PFC on an installation.
3. Draw, label, and explain the appropriate method of pulling a three-phase supply in three separate conduits.
4. Explain the significance of the term *percentage impedance*.
5. What is the abbreviation used for percentage impedance?
6. What is the formula used to calculate the full load current of a three-phase system?
7. Explain the relationship between the size of a conductor and the available short-circuit current?
8. Why is X/R ratio important in a short-circuit calculation?
9. Explain the relationship between the incremental distance of an installation and the time a circuit breaker takes to operate after a short circuit is created.
10. The service for an installation is 500 MCM × 2 per phase. The installation is located 500 ft from the supply transformer and is designed to supply load details as shown below. The supply transformer is rated at 300 kVA 208 V 3-phase and %Z = 5.5%. Taking into consideration X/R ratio between transformer and the installation, what is the value of the short-circuit current of the installation?
11. The service for an installation is 750 MCM × 2 per phase. The installation is located 500 ft from the supply transformer and is designed to supply load details as shown below. The supply transformer is rated at 300 kVA 208 V 3-phase, %Z = 5.5%. If the installation is located 100 ft from the supply transformer, taking into consideration X/R ratio between transformer and each of the installation, what is the value of the short-circuit current at each point of the installation?
12. Draw an illustration of a test to be done to determine the %Z of a transformer.

Test Your Knowledge Continued:

13. Carefully analyze the schematic diagram below and provide the kA ratings for Points C, D, E & F.

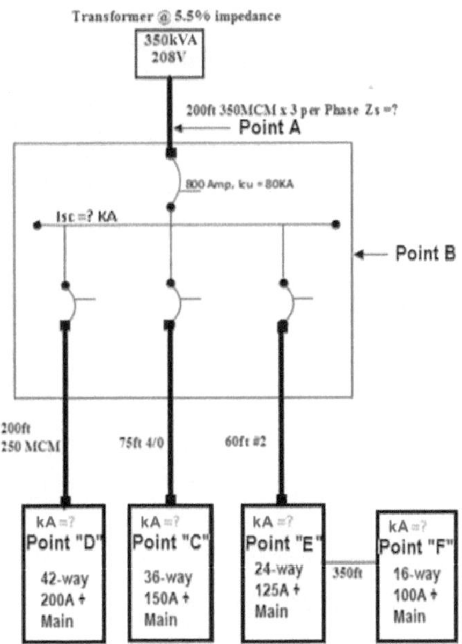

14. Using the kA ratings determined in question 13, complete the table below and determine the kA rating of each circuit breaker from Table 3.3.

Point	Design Current	Cal. PFC from question 13	kA rating selected from table 3.3	Quantity
A.B.D	200A	?	?	1
A.B.C	150A	?	?	1
A.B.E	125A	?	?	1
A.B.F	100A	?	?	1

Test Your Knowledge Continued:

15. Complete the table below from question 6 and table "F"

Point	Design Current	Prospective Fault Current in kA From Schematic	Breaker Type Selected From table "F"	Quantity
A.B.C	225 amp	?	?	1
A.B.D	175 amp	?	?	1
A.B.E	150 amp	?	?	1
A.B.F	25 amp	?	?	1

16. Identify and explain the areas marked with a question sign on the waveform below

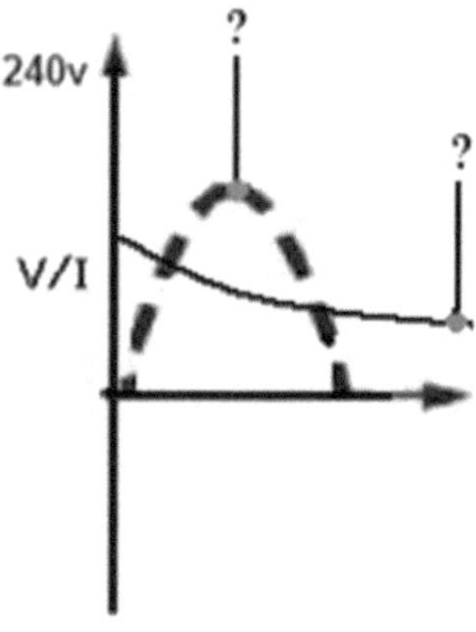

Bibliography

i. Cooper Bussmann. 2005. Electrical plan review. Retrieved in January 2013, from http://www.cooperindustries.com/content/dam/public/bussmann/Electrical/Resources/technical-literature/bus-ele-an-3015-electrical-plan-review.pdf

ii. Martino, F. J. Power Quality Drives LLC. 2001. Basic Short-circuit Capacity: Basic Calculations and Transformer Sizing. http://www.powerqualityanddrives.com/short_circuit_transformer/

iii. Roybal, D. D. June 2, 2006. Eaton Electrical. Cutler-Hammer Products. Primary and Secondary Distribution Systems. Retrieved in November 2010, from http://www.cfroundtable.org/meetings/060206/eaton.pdf

iv. Schneider Electric. Industries SAS. 06, 2009. *Medium Voltage Technical Guide*. Retrieved in June 2012, from https://www.google.com/url?sa=t&rct=j&q=&esrc=s&source=web&cd=10&ved=0CH8QFjAJ&url=http%3A%2F%2Fwww.ops-ecat.schneider-electric.com%2Fcut.CatalogueRetrieverServlet%2FCatalogueRetrieverServlet%3Ffct%3Dget_element%26env%3Dpublish%26scp_id%3DZ000%26el_typ%3Drendition%26cat_id%3DDesignerV4%26maj_v%3D2%26min_v%3D14%26nod_id%3D0000000013%26doc_id%3DH444155%26frm%3Dpdf%26usg%3D%26dwnl%3Dtrue&ei=dFebUtHiKo23kAefuYDIBQ&usg=AFQjCNGGjKdYL1BTRgKWpJZgz00hZTC1RQ&sig2=Z573ynP2MDBriUHtdF7ZMA

CHAPTER 5

Special Circuit Protection Devices

5.0 Introduction

Special circuit protective devices (SCPDs) are protective devices designed with microprocessors to detect and trip on micro faults in electrical systems. The most common type of SCPDs is the Ground fault circuit interrupter (GFCI). This device is fitted with a magnetic tripping device which activates within 4 to 35 mA.

5.1 GFCI Devices

GFCI, residual current devices (RCDs), and earth leakage current devices (ELCD) are devices which trip when leakage current is detected typically between 4-35 mA. These devices are intended to operate within 25-40 ms before an electric shock drives the heart into ventricular fibrillation. Ventricular fibrillation is a common cause of death due to electric shock. See Table 5.1 for effects of leakage current on humans.

The advantage of the GFCI device in comparison to the regular circuit breaker and outlet is that it has a quick disconnection time. This action is due to a built-in coil which is energized by a current ranging from 4 mA to 30 mA, depending on the sensitivity of the device. See Figure 5.1.

GFCIs help to protect against severe electric shock if a person comes in contact with live, exposed conductive parts. GFCI devices do not provide total immunity from electric shock. It is extremely essential to note that electric current does flow differently through a normal circuit load than it does when flowing through a person. Therefore, if a human body comes into contact with a live conductor protected by a GFCI, the body will act as a load and the device will operate within a time of 1 to 40 ms.

Table 5.1: Leakage current and effects on humans

Current in mA—Body to Ground	Body Effect
1 mA	Negligible sensation
10.5 mA	Finger muscles contract and fail to give up its grip
20 mA-30 mA	Restriction of breathing begins
> 50 mA	Current through body disorientates control signals to the heart (ventricular fibrillation <1%, but fatality is possible)

5.1.1 GFCI Principle of Operation

GFCI devices operate on a seesaw principle. This principle may be described as balancing current between the phase and the neutral conductor. It is a simple principle and a confirmation of one of Newton's laws which states, "For every action, there is an equal and opposite reaction." This means if 5 A is flowing through a phase conductor, on principle, 5 A should be returned on the neutral conductor as shown in Figure 5.1 (A).

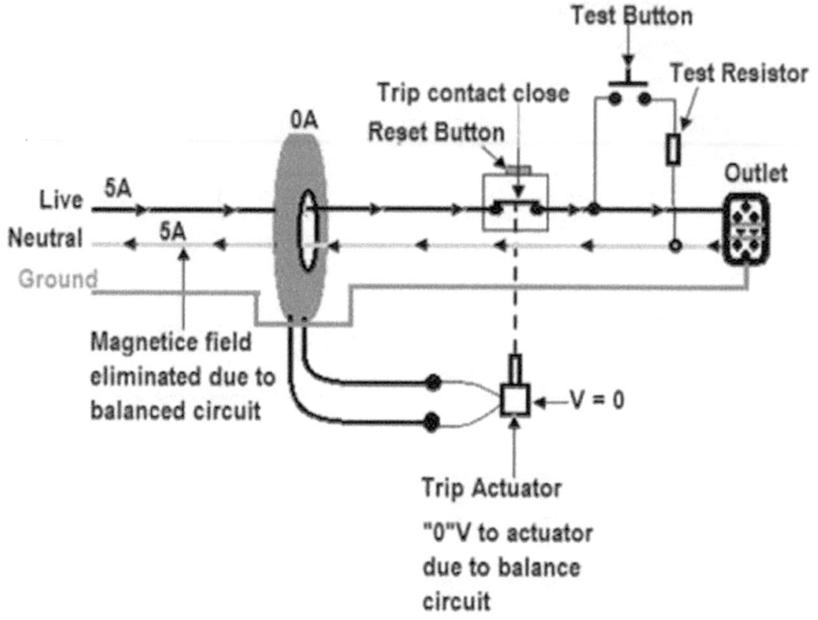

(A) *Normal operating GFCI receptacle*

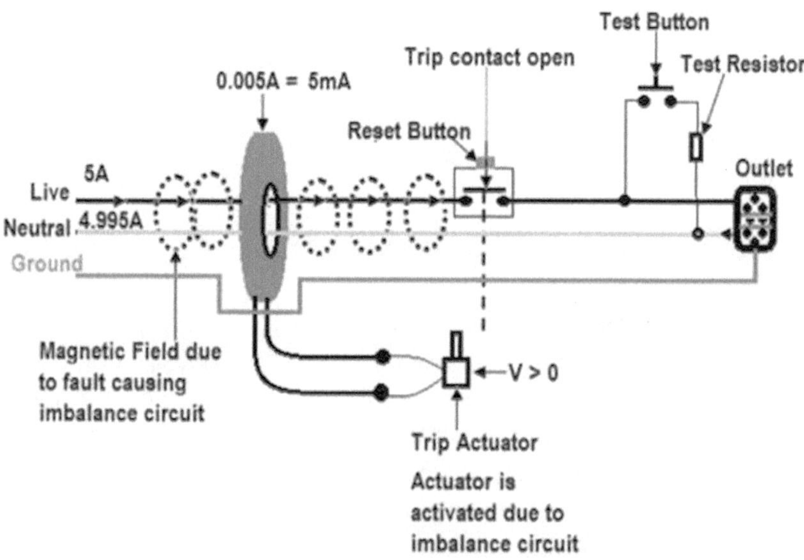

(B) *Typical GFCI outlet trips due to a fault*

Figure 5.1: *Typical internal wiring and operation of a GFCI receptacle*

Fault current on appliances is sometimes not enough to operate an ordinary circuit breaker unless there is a direct short to ground on the appliance; however, appliances supplied by GFCI receptacles will operate as a shunt resistor, creating a difference in the amount of current returning through the phase or neutral conductor as shown in Figure 5.1 (B). Fault current between 10 to 50 mA is fatal to the human body, and it is for this reason that GFCI devices are designed to trip when it detects 5 to 10 mA difference between the current-carrying conductors. The trip limits of GFCI devices reduce the possibility of fatal fault current developing on the protected installation.

How does the human body cause a GFCI device to trip? The body will act as a shunt resistor by diverting a portion of the current to the general mass of earth, and this will create an imbalance in the device, thus causing the device to trip.

5.1.2 Water Bucket Illustration of the Operation of the GFCI

The water bucket illustration as shown in Figure 5.2 simulates how the GFCI operates. This can be viewed as balancing a seesaw with two buckets of water.

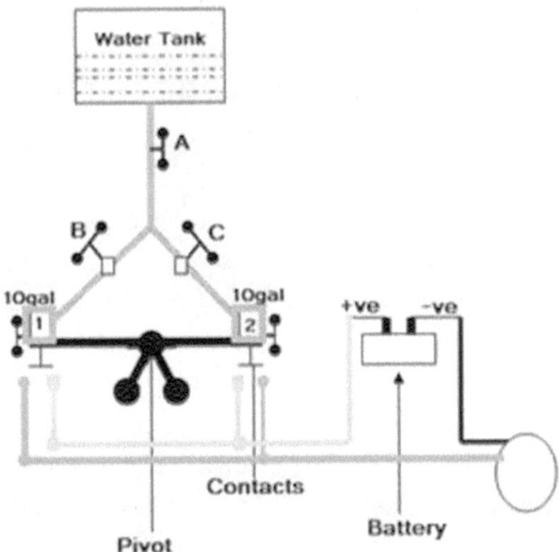

Figure 5.2: *GFCI and balancing principle*

Two 10 gallon water tanks (tank 1 and tank 2) fitted with drain valves are placed on each side of a balancing board. Each tank is filled with 10 gallons of water from the supply tank. Beneath each end of the board is a normally open contact. Both contacts are attached to a 12 V battery source and a 12 V indicator. Valves *B* and *C* on the supply of tanks 1 and 2 must be placed in the open position while the valve *A* remains closed.

The 12 V indicator will remain off *once tank 1 and tank 2 remain balanced.* The distance between each contact represents a short-circuit current of 5 mA. If the drain valve on tank 1 opens slightly, water will be released creating an imbalance between the two tanks. Tank 2 will then become heavier, which will result in the tank moving downward, closing the contact point below tank 2. This sends power to the lamp, causing it to be lit.

If the process is reversed, the lamp will be lit, and the light will remain on until the tanks are refilled and both tanks have the same volume of water.

5.2 Practical GFCI Circuit Breaker

The outline of the GFCI shown in Figure 5.3 and the typical construction diagram (Figure 5.4) are depicting that of a GFCI circuit breaker. The principle of operation of this circuit breaker is the same as that of a GFCI receptacle or outlet. The GFCI circuit breaker is an evolution to the earth leakage circuit breaker (ELCB), which operates on the principle of the earth becoming energized. GFCI circuit breakers operate on the principle of unbalancing between the phase and neutral or phase and phase circuit conductor. This imbalanced condition may be due to a person unintentionally coming in contact with a GFCI-protected conductor. Unintentional contact will cause the circuit to become imbalanced with currents which exceed the safety limit threshold of 5 mA between phases and neutral for single-phase 120 V devices or 5 mA between phases for 220 V circuits. A GFCI circuit breaker will not prevent an electrical shock if unintentional contact is made with a live conductor from GFCI-protected circuits. A

GFCI circuit breaker will reduce the duration of the shock, thereby protecting life. Figure 5.4 shows the construction details of the GFCI circuit breaker.

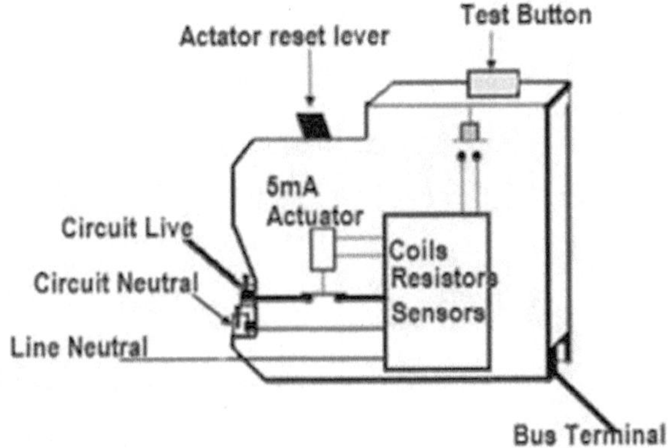

Figure 5.3: *Outline of a typical GFCI circuit breaker*

A GFCI circuit breaker is used to protect circuits in wet or damp locations, for example, patios, garages, courtyard, etc. Open areas of schools and churches where receptacles are required should also be fitted with GFCI circuit breakers. This will reduce the risk of mischievous children or persons being electrocuted.

For general household purposes, GFCI circuit breakers can be used for circuits which supply countertop receptacles, bathroom receptacles, and outdoor circuits.

GFCI devices are prone to a phenomenon known as nuisance tripping. This phenomenon is defined as intermittent operation due to sensitivity. A GFCI device is designed to operate if there is an unintentional contact with a person or animal; however, the following will trigger an operation of the device or cause nuisance tripping:

1. Moisture
2. Poor installation of device
3. Defective equipment or appliance
4. Defective GFCI device
5. GFCI in direct contact with a splash or rain water

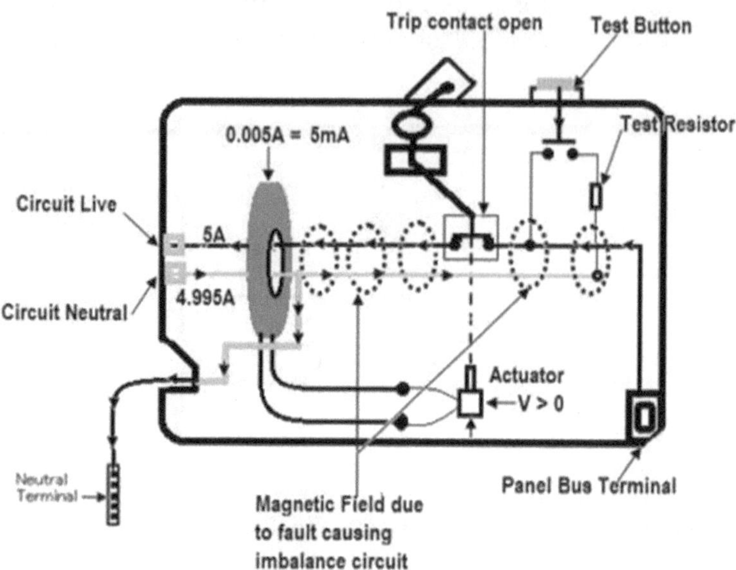

Figure 5.4: *Typical construction of a GFCI circuit breaker*

5.3 Arc Fault Circuit Interrupter (AFCI)

A typical AFCI shown in Figure 5.5 monitors two significant types of fault signatures. These are parallel and series arc signatures (PAS & SAS). The word *signature* is used to describe the uniqueness of the occurrence of these events. PAS and SAS are faults which cannot be detected by the regular GFCI or circuit breakers. These faults produce low harmonic waveforms which are only detected by electronics fitted inside the protective device.

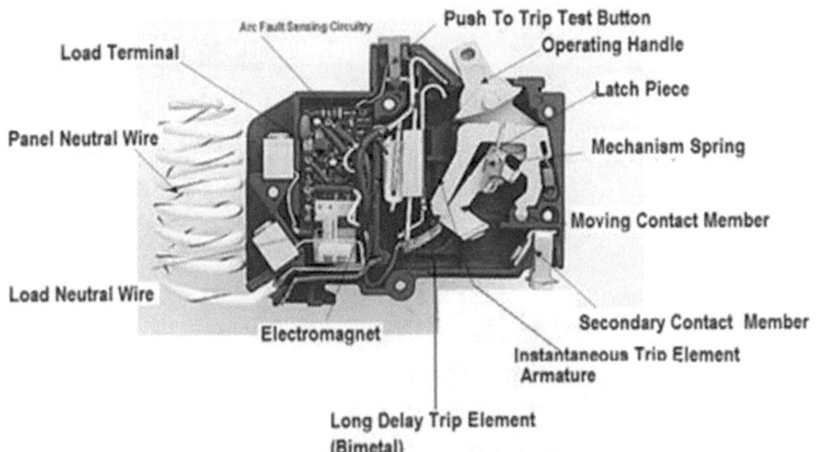

(Adapted from Eaton Documentation—Technical Papers. n.d.)
Figure 5.5: *Arc fault circuit interrupter*

The arc fault circuit interrupter is quite similar to a GFCI in appearance, but it contains a high-tech microprocessor that constantly monitors the current waveform as electricity passes through the breaker. The control circuit in the AFCI is capable of "tripping" the breaker when the microprocessor detects certain current waveforms that are associated with line-to-neutral arcing.

A GFCI can protect personnel from ground faults, but if the fault is line-to-neutral, the same current passes through both of the intended conductors of the circuit, and there is no magnetic imbalance. The AFCI acts like an electronic oscilloscope programmed to open the circuit breaker when it identifies particular current waveforms as shown in Figures 5.6 and 5.7.

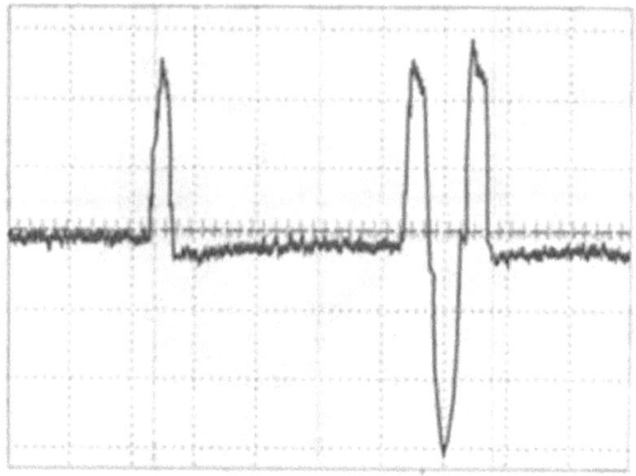

(Adapted from Electronics Principles and Applications Seventh Edition, 2008)

Figure 5.6: *Typical harmonics showing high arc signatures*

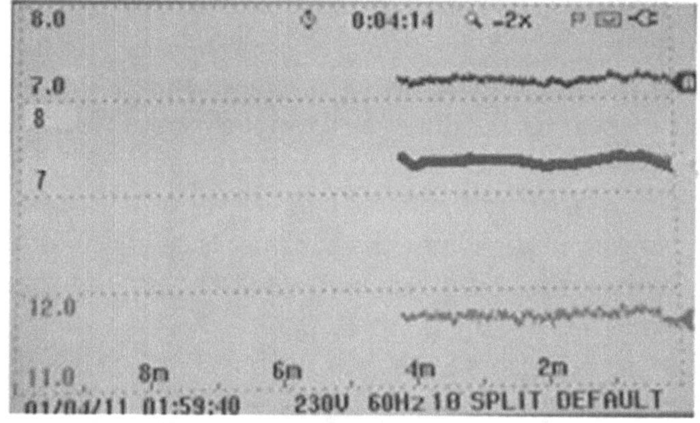

Figure 5.7: *Normal harmonic levels showing no arc signatures*

5.4 Parallel Arc Signature (PAS)

Parallel arc faults are indirect contact fault. These occur when the insulation of parallel conductors (live and neutral) in the same conduit or shroud is damaged and is carrying a full load or high current. A parallel fault occurs when current passing through a conductor breaks the air gap between live and neutral as shown in Figure 5.8, This creates an arc which generates heat up to 6000° C. The heat created by this fault causes the insulation of other cables to break down, further causing direct short circuits and significant damage to all cables within the conduits or shroud.

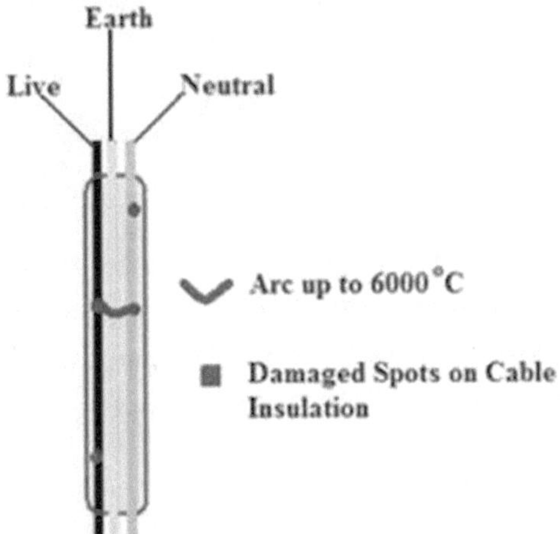

Figure 5.8: *Parallel arc signature*

5.5 Series Arc Signatures (SAS)

Series arc faults occur in individual live or neutral conductors of flexible cords supplying appliances such as electric irons, telephone chargers, laptop adaptors, etc. Series arc faults are frequent occurrences and are referred to as cable break. After this break occurs, carbon is built up between the break points, producing minute conductivity and arcing between these points. Arcing will occur between these two points until the gaps become wider and the built-up carbon is burnt out. Figure 5.9 shows a typical series arc fault.

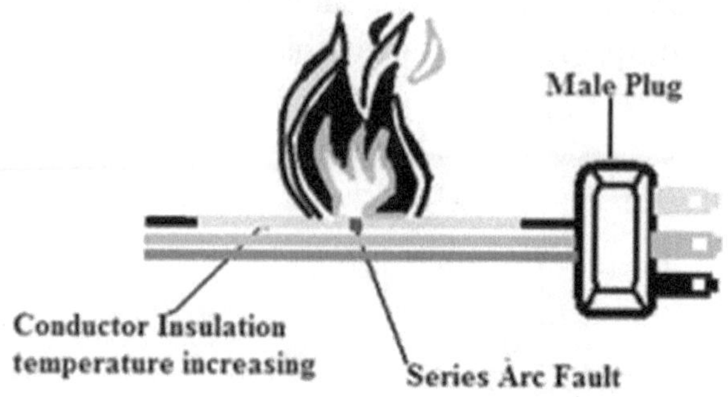

(Adapted from Electronics Principles and Applications Seventh Edition, 2008)

Figure 5.9: *Series arc fault (cable break)*

5.6 Relay Protection

Protective relays are devices which monitor current, voltage, and frequency for manufacturing plant equipment, distribution, and transmission lines. These protective devices are extremely reliable in isolating equipment when their predetermined parameters are breached.

While electromagnetic or thermal circuit breakers are simple protective devices used for LV installations, protection for HV systems is provided by air blast, oil, SF6, or vacuum circuit breakers. These HV circuit breakers employ trip coils activated by protective relays. These relays monitor the current entering and leaving the distribution lines which they protect.

Overcurrent protective relays fall into one of the following categories:

1. Electromechanical/electromagnetic
2. Solid state
3. Computer-based

The characteristics of relays are as follows:

1. Instantaneous
2. Definite time
3. Inverse time

5.6.1 Inverse Definite Minimum Time Lag (IDMTL) Relay

IDMTL relays are the most common type of electromagnetic relay used to provide protection for the HV system. The construction of this device is similar to a kilowatt-hour meter and operates on the principle of induction. IDMTL relays are not directly coupled to the lines they protect; they rely on current transformer (CT) and voltage transformer (VT) coupled to these lines for monitoring current and voltage parameters.

5.7 Motor Overload Protection

High inductive load such as motors should be fitted with magnetic starters. Motor starters generally have the following in their framework:

1. Contactors
2. Overload protection
3. Phase monitor (optional)
4. Stop-start push button
5. Auxiliary contacts

Motor or overload protection is designed for protecting rotating magnetic machines or any other such type magnetizing equipment from over-magnetizing current. Over-magnetizing current is classified as a fault current which is sometimes significantly high in motors.

Overload protection fitted to motor contactors is either electromechanical or solid state. Based on the type of protection required, one of the solid state overload relays shown in Figures 5.10 to 5.12 may be used.

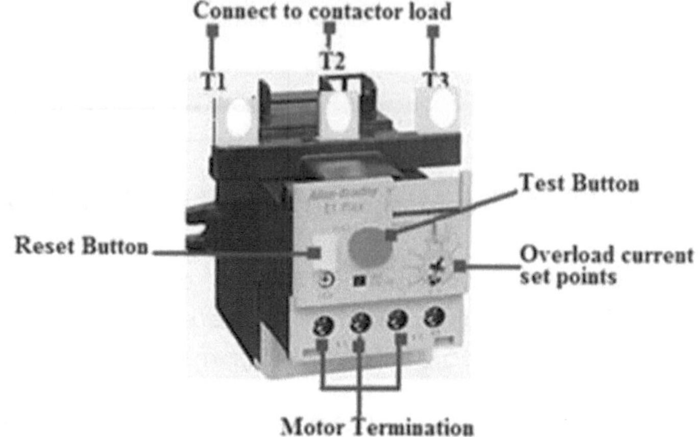

(Adapted from Rockwell Automation, Allen-Bradley, 2013)
Figure 5.10: *Solid state thermal overload relay*

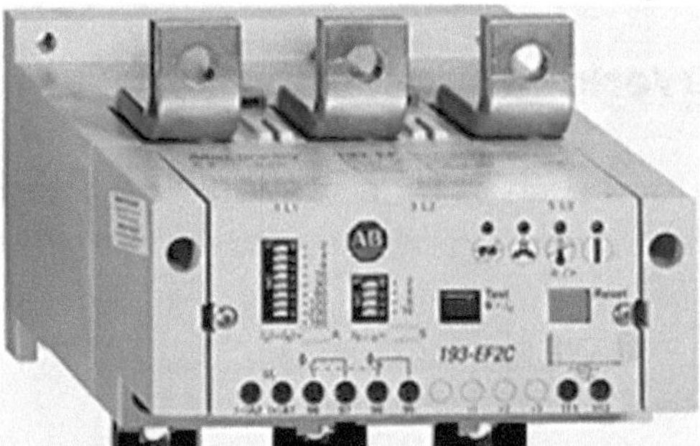

(Adapted from Rockwell Automation, Allen-Bradley, 2013)
Figure 5.11: *Solid state overload relay*

(Adapted from Thermal Overload Relay, n.d.)
Figure 5.12: *Thermal overloads*

5.8 Understanding Motor Control Circuits

Protective devices for motors are designed to activate or deactivate the coils of contactors or relays in their control circuits. These control circuits are fitted on equipment to protect and manipulate operations.

Controlling electrical motor can be done by using one of the following methods:

1. Manual (stop, start, reverse, and forward)
2. Automatic (stop, start, reverse, forward, limit switches, float switches, level sensors, photo sensors, and so on)
3. Reduced voltage
4. Variable speed drive
5. Intelligent controllers

5.8.1 Installing Protective Devices

Common to each method of motor control and the wiring arrangement of their protective device is the series connection between the contactor coil and the following devices:

1. Stop switches
2. Emergency switches
3. Limit switches
4. Sensors
5. Float switches
6. Level sensors

Figures 5.14 - 5.24 shows connections for installation of protective devices in motor control circuits.

5.8.2 Relay or Contactor Terminals Identification

Relay contactor terminals are the points where controlling and protective conductors are terminated. These terminals are labeled in pairs, for example, 1 and 2, 3 and 4, and 5 and 6 or *T1* and

*T*2, *T*3 and *T*4, *T*5 and *T*6, and so on, as shown in Figure 5.13. Terminal numbing or lettering is designated numbers and may vary depending on manufacturers.

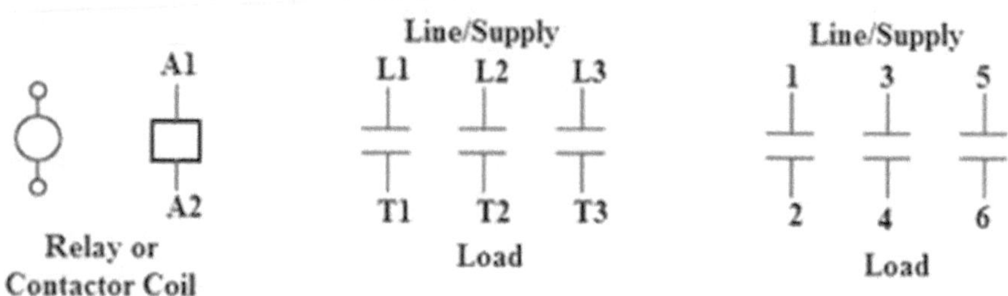

Figure 5.13: *Typical identification of starter terminals*

5.8.3 Dry Contacts

In addition to the normal power transferring contacts in a control system is the inclusion of auxiliary contacts, called dry or no volt contacts. These contacts are classified as normally open and normally closed. These contacts are common to most control wiring systems. The number of dry contact in a circuit varies with circuit complexity. Their abbreviations and applications are as follows:

1. Holding contacts (N/C)
2. Running pilot light (N/O)
3. Stop pilot light (N/O)
4. Overload trip contacts (N/C)

Auxiliary contacts are normally located on the side or the top of the starter and are normally labeled as shown in Table 5.2. This Table 5.2 shows typical terminal numbering/lettering or symbol and their definitions.

Table 5.2: Terminals and contact identification

Terminal Numbering/ Lettering or Symbol		Definition
L1	1	First line of the power source
L2	3	Second line of the power source
L3	5	Third line of the power source
T1	2	First line out of the contactor to the motor terminal
T2	4	Second line out of the contactor to motor terminal
T3	6	Third line out of the contactor to motor terminal
1-2, 13-14, 53-54	-||-	Normally open *N/O* contact (closes when the relay energizes)
95-96	-|/|-	Normally closed NC contact (opens when the thermal overloads trip if associated with the overload relay or block)
-|/|-		Normally closed *N/C* contact (opens when the relay energizes if associated or linked with the relay coil as auxiliary circuit)
M		Relay/contactor coil

5.8.4 Motor Control Wiring

Understanding the basics of control wiring is extremely vital before attempting to wire a control circuit or an overload protection for a motor. There are two methods of wiring a motor control. They are as follows:

1. 2-wire control
2. 3-wire control

5.8.4.1 Two-wire (2-wire) Control

Two-wire (2-wire) control circuit is defined as two wires connected in series to a device (float, limit levels, etc.) to the line side of a starter or contactor coil. This type of wiring is used where an unmonitored control system is required. For example, if a power failure occurs when equipment with 2-wire control wiring is in operation, the system will reset itself as soon as the power is restored. Figure 5.14 shows the ladder diagram for wiring a 2-wire control.

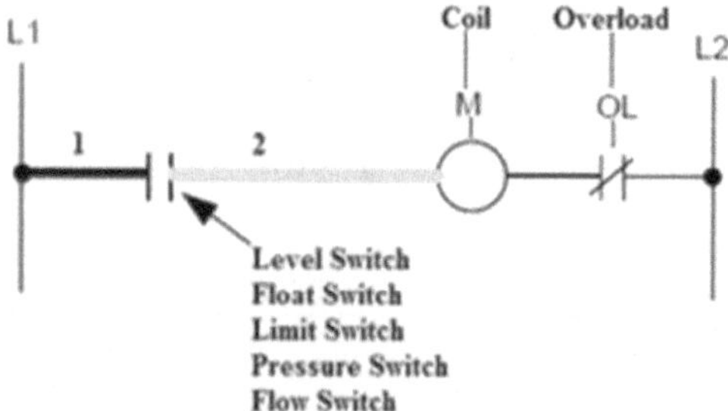

(Adapted from www.automationdirect.com.
How to Wire a Motor Starter, October 2, 2005)
Figure 5.14: *Two-wire (2-wire) control ladder diagram*

5.8.4.2 Three-wire (3-wire) Control

A three-wire (3-wire) control circuit is defined as three wires connected to a momentary switch such as a push button (start, stop, jog, forward, and reverse) or any other similar devices. Momentary switches in a three-wire control circuit are connected in series parallel with the line side of a starter or contactor coil. This type of wiring is designed to protect against an unexpected start-up of a machine after a power failure has occurred. See Figures 5.15, 5.16 and 5.17.

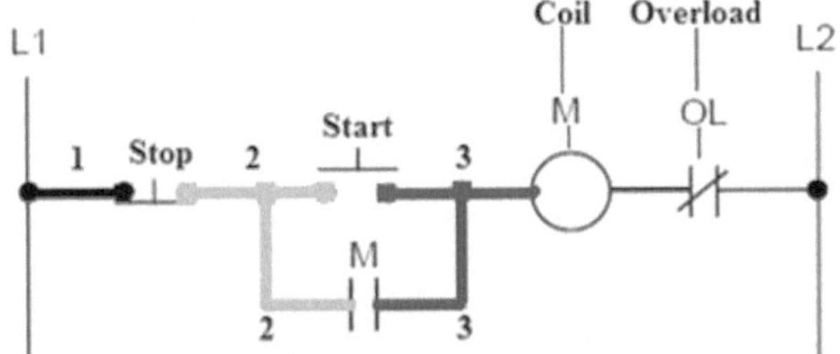

(Adapted from www.automationdirect.com.
How to Wire a Motor Starter, October 2, 2005)

Figure 5.15: *Three-wire (3-wire) control ladder diagram*

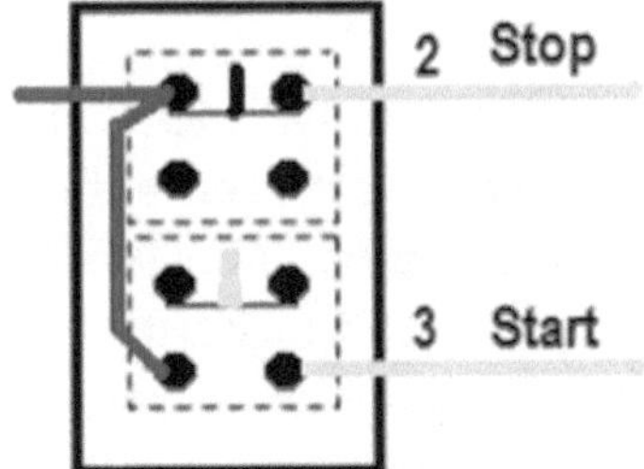

Figure 5.16: *Typical three-wire control station*

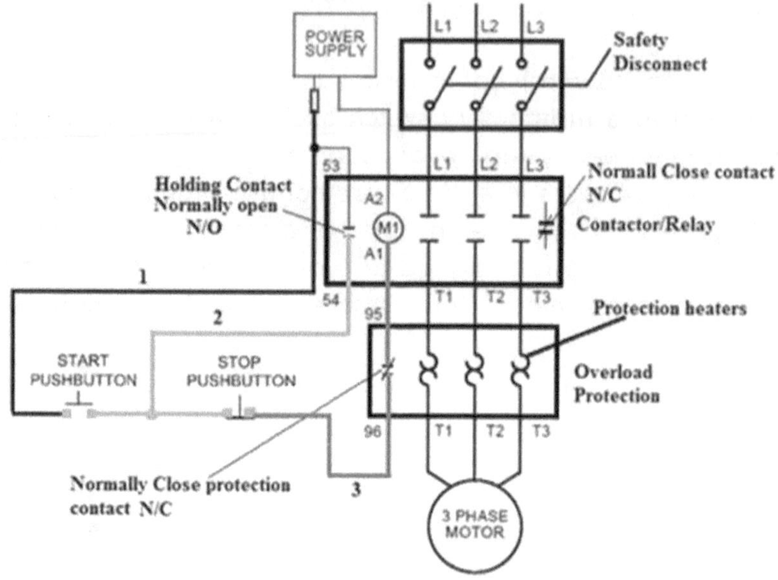

(Adapted from www.automationdirect.com.
How to Wire a Motor Starter, October 2, 2005)

Figure 5.17: *Typical 3-wire control for a three-phase direct online start*

5.8.4.3 Three-wire (3-wire) Motor Control Including Jog Control

Jog controls are used in industrial and commercial environments where motor application requires movement which is not continuous. *Jogging* is sometimes referred to as inching forward or backward. This type of wiring is done at the control station for machinery, conveyor belts and other devices which require micro management of their movement. A typical control circuits is represented by the ladder diagram shown in Figure 5.18. Figure 5.19 provides terminals for a typical start, stop, jog control. Bypassing the holding contact *M* shown in the ladder diagram prevents the coil from connecting to a permanent supply once the push button is pressed.

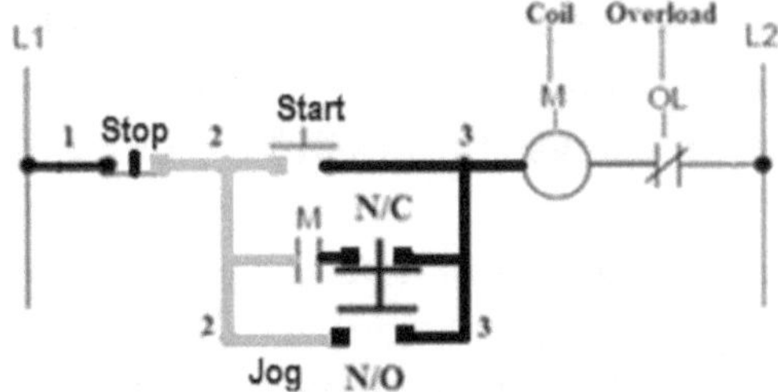

(Adapted from www.automationdirect.com. How to Wire a Motor Starter, October 2, 2005)

Figure 5.18: *Three-wire (3-wire) ladder diagram including jog control*

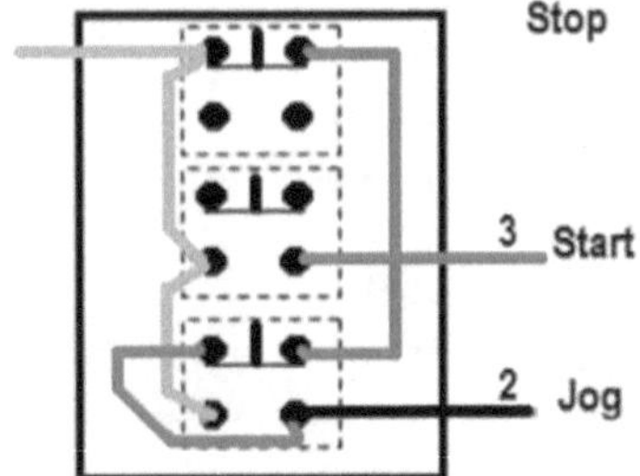

Figure 5.19 *Typical three-wire control station including jog*

5.9 Forward/Reverse/Jog Control for Three-phase Motors

Three phase induction motors posses the advantage of rotating in both direction with little effort. The direction of rotation of these motors can be changed by simply interchanging any two of the three phases, supplying the machine or motor This characteristic is exploited in forward, reverse and jog controls for these motors.

Phases can be interchanged physically by interchanging L1 with L2 or L3 or by using two contactors M1 and M2 with electrical or mechanical interlocks. Interlocks in forward reverse controls are used to prevent contactor number 1 (M1) and contactor number 2 (M2) to operate simultaneously. In addition, L1 for coil M1 is passed through a normally close (N/C) contact on contactor number 2 (M2); L2 for coil M2 is passed through a normally close

(N/C) contact on contactor number 1 (M1). Therefore, if a machine is running in the forward direction, the reverse contactor will not energize without the forward contactor de-energized.

Short-circuit current between two phases in this type control configuration could be catastrophic; therefore, care must be taken in the wiring of these type of motor controls. Figures 5.20 to Figure 5.24 show the typical ladder diagrams and wiring methods, including motor protection for forward/reverse control.

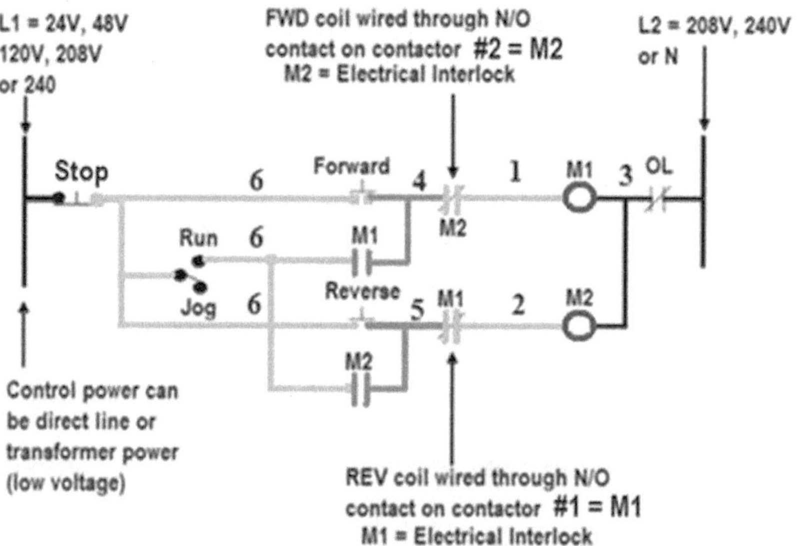

(Adapted from www.automationdirect.com. How to Wire a Motor Starter, October 2, 2005)

Figure 5.20: *Typical forward/reverse ladder diagram with electrical interlock*

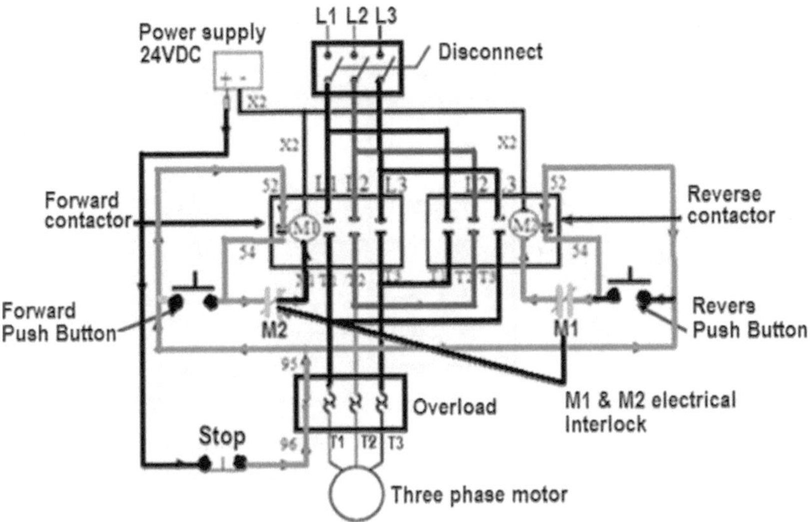

(Adapted from www.automationdirect.com. How to Wire a Motor Starter, October 2, 2005)

Figure 5.21: *Typical forward/reverse wiring complete with electrical interlock*

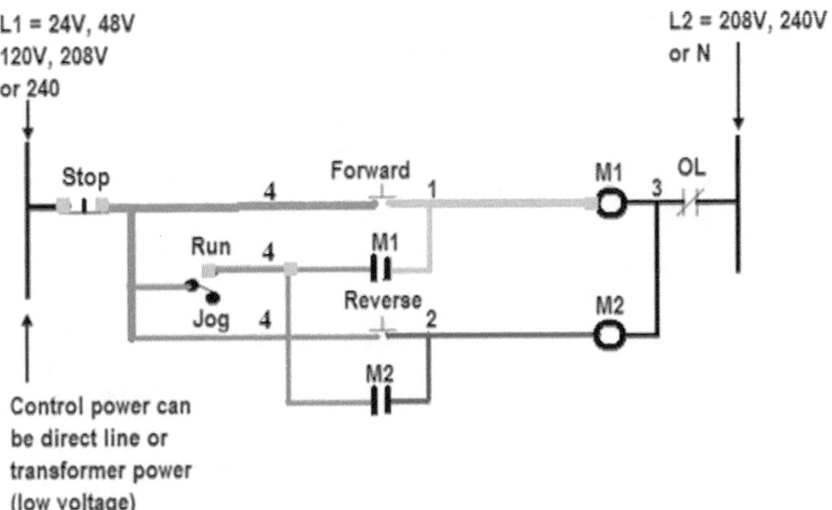

Figure 5.22: *Typical forward/reverse ladder diagram with mechanical interlock*

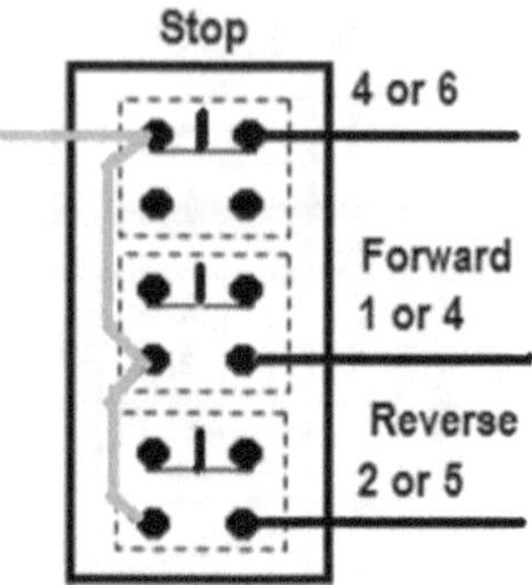

Figure 5.23: *Typical forward/reverse control station. Numbering is based on the type of associated interlocking system used*

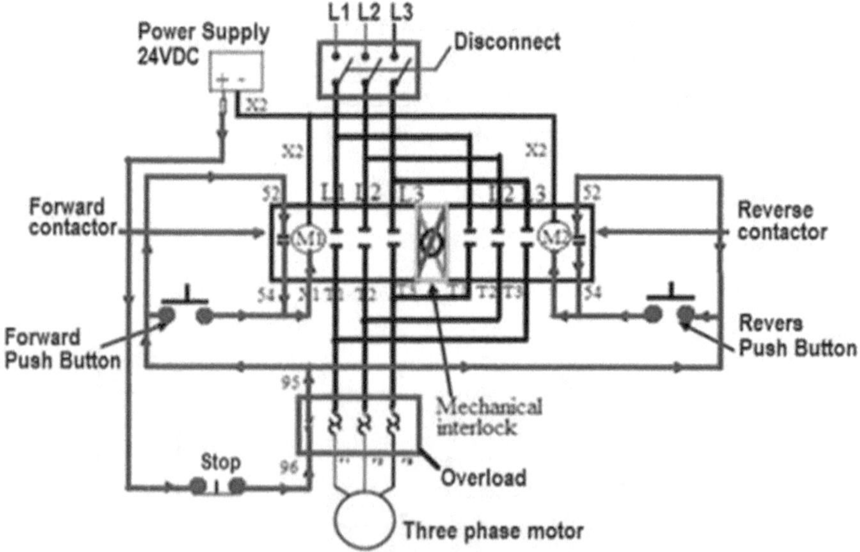

(Adapted from www.automationdirect.com. How to Wire a Motor Starter, October 2, 2005)

Figure 5.24: *Typical forward/reverse wiring complete with mechanical interlock*

5.10.1 Typical Overload Load Wiring

Common to all motor control wiring, the main protection is the motor overload shown in Figures 5.21 and 5.24. Figure 5.25 shows the motor terminals *T*1, *T*2, and *T*3, passing through the heaters located at the top section of the overload block. The main control wire through the N/C contact is located at the bottom section of this overload block. The location of normally close contacts (N/C) or terminal varies, depending on the type and brand of overload used. Figure 5.25 also shows a typical internal mechanism of an overload. This may also vary, depending on the type and brand of to the overload.

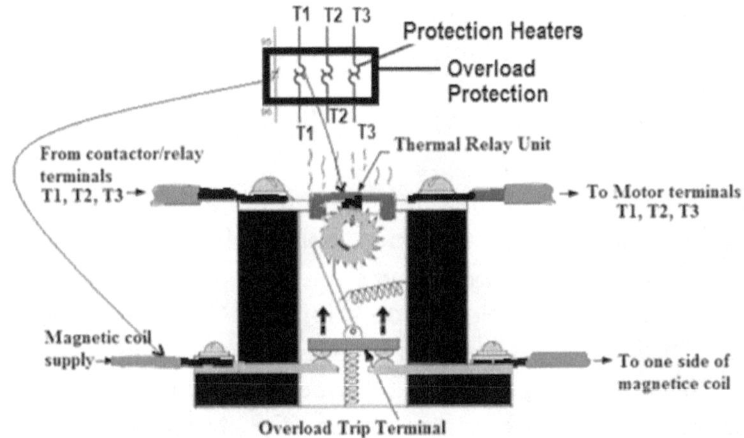

Adapted from Overload Relays and Thermal Unit Selection, 2013)

Figure 5.25: *A typical overload block*

5.10 Conclusion

Special protective devices prevent electrocution and protect property if a live conductor should come in contact with moisture or wet surfaces. These devices quickly disconnect circuits when faults such as arc signatures occurs. Fault currents may be as low as 5mA, however, fault current of this nature is not enough to activate regular circuit breakers. As a consequence, it is recommended that sensitive equipment should be protected by GFCI circuit breakers or plugged into GFCI receptacles.

Fault current generated on a motor control circuit can be catastrophic. Understanding motor control wiring and the coordination between the control station (stop, start, jog, pilots, etc.), coils, overload protection, and main supplies help reduce down time if a fault occur and reduce the risk of catastrophic failures caused by incorrect wiring and high fault currents.

5.11 Test Your Knowledge

1. State the principle of operation of a GFCI device.
2. What are the effects of 10 mA of current on the human body?
3. Explain why a GFCI device will not prevent an electric shock.
4. Draw and label the internal wiring of a GFCI receptacle.
5. What is the difference between the ELCB and the GFCI?
6. What is nuisance tripping?
7. State the principle of operation of a GFCI receptacle.
8. List five critical areas where GFCI receptacles are used.
9. What is the operating current of a GFCI receptacle?
10. What is the operating time of a GFCI receptacle?
11. Explain the difference between an arc fault circuit breaker and a ground fault circuit breaker.
12. What is meant by the terms *parallel arc signature* and *series arc signatures*?
13. Draw and label examples of parallel arc signature and series arc signatures.
14. Sketch the waveforms for parallel arc signature and SAS.
15. Draw and label a complete control diagram for direct online starting.
16. What is the importance of wiring the power supply through the overload block?
17. What is the key feature when wiring a jog control?
18. Draw and label a complete wiring diagram of a forward/ reverse control system.

Bibliography

i. Underwriters Laboratories Inc. 2001. Arc fault circuit interrupter (AFCI) FACT sheet. Retrieved in May 2011, from http://www.bentcreekinspection.com/Library/GFCI%20Circuit Interrupter info.pdf

ii. Electrical Advice. More information about GFCIs and AFCIs. n.d. http://www.nfphampden.com/gfc_afci.pdf

iii. Schuler, C. A. 2008. Electronics: Principles and Applications Seventh ed. Arc Fault Circuit Interrupters. Retrieved on July 3, 2013, from 2008 McGraw-Hill

iv. GE Imagination at Work. February 2009. Keep Him at the Station—Combination Arc Fault Circuit Interrupters. Retrieved on July 3, 2013, from http://www.geindustrial.com/publibrary/checkout/DEA-437A?TNR=Brochures|DEA-437A|generic

v. Eaton. May 2013. Molded Case Circuit Breakers and Enclosures. Arc Fault Circuit Breakers. Retrieved on September 20, 2013, from http://www.eaton.com/Eaton/ProductsServices/Electrical/ProductsandServices/Residential/LoadcentersandCircuitBreakers/CHCircuitBreakers/CHArcFault/index.htm

vi. Goldwasser S. Ground Fault Circuit Interrupters. 1998-2013. Retrieved on September 20, 2013, from http://www.rhtubs.com/GFCI/GFCI.htm

vii. Rockwell Automation, Allen-Bradly. 2013. Solid-state overload relays. Retrieved on July 17, 2013, from: http://ab.rockwellautomation.com/Circuit-and-Load-Protection/Motor-Protection/Solid-State-Overload-Relays

viii. Schneider Electric. June 2003. *HV Training Manual*. www.automationdirect.com.How to Wire a Motor Starter. October 2, 2005. Retrieved on July 17, 2013, from http://support.automationdirect.com/docs/an-mc-004.pdf

ix. EC&M Electrical Construction and Maintenance. 2013. Series and Parallel Arcing. Retrieved on July 27, 2013, from http://ecmweb.com/mag/electric_basics_arcfault_protection/

x. Square D, Group Schnieder. 1998. Overload Relays and Thermal Unit Selection. Retrieved on July 18, 2013, from http://www.hagemeyerna.com/getdoc/ff24e3bf-7ac7-4e0d-8790-dcbefd41aed8/SquareD-Overload-and-Thermal-Selector.aspx

xi. Thermal Overload Relays n.d. Retrieved on July 17, 2013, from http://www.lntebg.com/brochures/switchgear/07-Relays1.pdf

CHAPTER 6

Grounding Systems and Functions

6.0 Introduction

Grounding systems are designed to protect life and property by providing an effective path for unwanted current and voltage to be dissipated to the general mass of earth. Ground current-monitoring devices constitute a significant element in electrical systems for providing protection. The systems establish set points which are used to remove undesirable currents and voltages quickly when the earth leakage set points are breached. Electronic systems such as computers or computer-based controls are mostly affected by ground faults or high harmonic rich currents which are called noise. Supplementary grounding systems are sometimes employed to achieve desired connection with the general mass of the earth.

6.1 Electrical Noise

Unwanted currents and voltages which cause interference on electrical systems are called noise. Noise developed on an electrical system could be as a result of currents and voltages straying from their point of origin to other parts of an electrical system or appliances via shielded or unshielded conductors. Stray current and voltage can also be due to harmonics generated by electrical devices such as solenoids, relays, motors, power lines, ignition systems, and so on. Figure 6.1 demonstrates a magnetic field linking a shielded cable and inducing current in the shield which flows to earth.

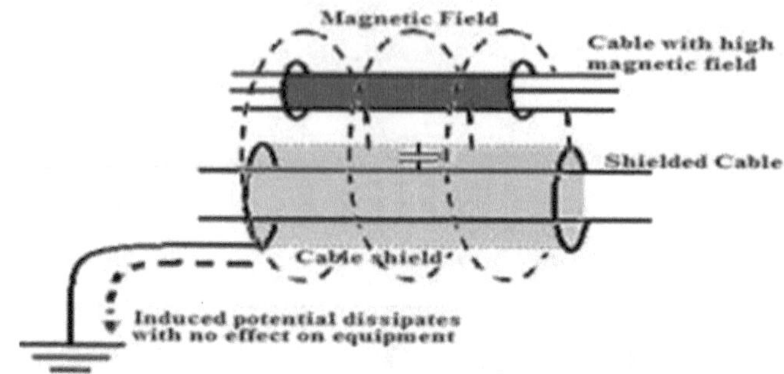

Figure 6.1: *Shielded cable running alongside power cables with high magnetic field*

Noise can present negative effects on sensitive systems. Maples System Inc. Technical Notes 1027, on OIT (output/input terminal) Grounding Wiring and Electrical Noise Reduction, attributes some malfunctioning of electrical devices, equipment, and appliances by the presence of high noise interference to "display mysteriously clearing by themselves, or keyboards 'locking up' and stop responding to operator input."

Maples System Inc. Technical Notes 1027, further states, "The system may respond to a reset, or the system may have to be turned off physically and then back on again, at which point it starts to operate as if nothing had happened." Furthermore, "There may be an obvious cause, such as an electrostatic 'spark' from someone's fingertip to a keyboard or the interference may occur every time a particular motor or solenoid is turned on or off. There may be no obvious cause, and there may be no obvious cause, and there may be no action that the operator could perform to recreate the fault. However, a few minutes, or a few hours, or a few days later it happens again."

Figure 6.2 shows grounding connections that will create a circulating current path.

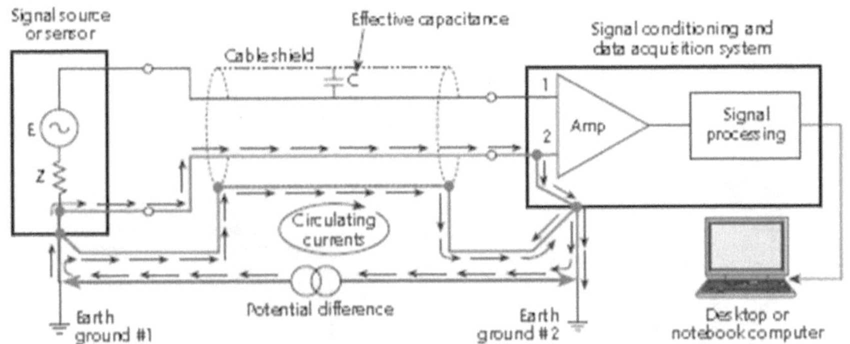

(Source: TECHTIP 60201 How to Reduce System Noise, n.d.)

Figure 6.2: *Potential difference created between equipment coupled with low signal shielded cable*

In this system, take notice of the connections and the arrow lines between the shielded cable, earth #1 and earth #2. This type of cable connection will cause signal interference in sensitive equipment.

Methods of connection and types of shielded cable to reduce noise interference in control systems are as follows:

1. Connect all cable shields to chassis ground.
2. Connect all grounding or earthing conductors to grounding terminals.
3. Never connect the shield of a cable and the grounding or earthing conductors together.
4. Use cables shielded with copper, aluminum, or tin.
5. Implement shield in areas where high magnetic fields are present, non-ferrous materials can be used to provide shielding.

When connecting shielded cables between two separate power source equipment, the shield of each cable should be connected to the ground at one end, preferably the end where the signal is generated. This will eliminate circulating currents, thus making the shield of the cable effective. In Figure 6.3, the path for circulating currents has been eliminated.

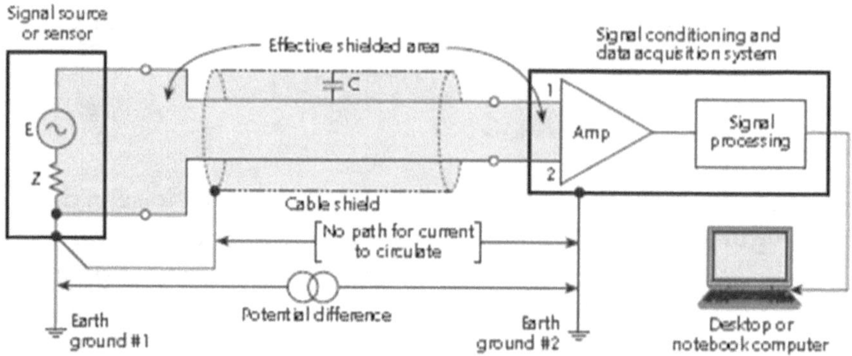

(Source: TECHTIP 60201 How to Reduce System Noise, n.d.)

Figure 6.3: *Effective grounding: Shield of the cable connected at the source only*

A well-organized grounding and shielding system as shown in Figure 6.4 will provide a clean path for noise (unwanted currents and voltages) to dissipate to the general mass of earth. This will also eliminate system interference by stray currents and voltages. Additionally, an organized grounding and shield system will provide protection from a combination of the following:

1. Phase-to-ground or neutral fault at any point of the installation
2. Transients
3. Lightning
4. Harmonics current

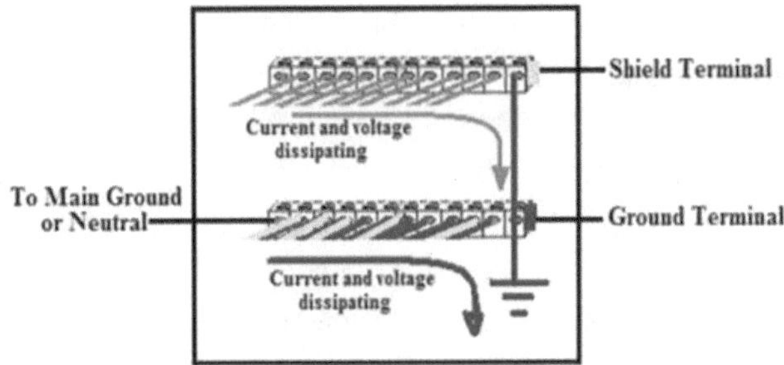

Figure 6.4: *Connecting equipment ground and shields of cables to eliminate system interference*

6.2 Grounding Path

Creating a perfect grounding path is crucial in any electrical installation. It is also essential to achieve perfect connectivity between the general mass of earth and the earth electrode; this connectivity will provide quick conductivity of fault current to the general mass of earth.

The NEC refers to quality connectivity to the general mass of earth, as achieving a resistance of 25 Ω (ohms) between the earth electrode and the general mass of earth using an earth resistance tester. Instruments that could be used to determine this resistance may include *three-point ground tester, four-point ground tester,* or *clamp-on ground tester.*

In the event of a short circuit existing between phase-to-neutral or phase-to-earth, the neutral conductor carries the fault current and acts as a transformer star point. This eliminates the myth that short-circuit currents are carried by the earth electrode and general mass of earth to the transformer.

The IEE *On-site Guide* of the IEE Regulation, page 133, gives guidance on effective electrode resistance as follows:

> The earth electrode resistance of TT installations must be measured, and normally a Residual Current Device (RCD) is required.

> For reliability in service the resistance of any earth electrode should be below 200 Ω.

The NEC code states:

> A single electrode consisting of a rod, pipe or plate that does not have a resistance to ground of 25 Ω or less shall be augmented by one additional electrode of the types specified in section 250-50 and 250-52.

6.3 Ground Resistance for Domestic, Transmission, Radio, and Television Antennas

In the foregoing quotes on earth resistance, the NEC makes reference to 25 Ω as a base reading, and the IEE refers to 200 Ω as a base resistance. The average of these resistances (112.5 Ω) is considered adequate for providing protection for an installation. It may be necessary to provide an additional electrode if the soil condition is poor.

For domestic installations, a ground resistance of 25 Ω should be the target; however, depending on the terrain and soil quality, a single earth electrode which produces a resistance of 200 Ω or below is acceptable. If it is absolutely necessary to satisfy the two codes, this electrode should be augmented by one or more electrodes of similar length and spaced at ten feet intervals as shown in Figure 6.5. In addition, the I.E.E ON-SITE GUIDE of the I.E.E Regulation: (Page 133) gives guidance on the resistance an effective earth electrode, not greater than 200 ohms.

Earth resistance for transmission and distribution lines should be below 10 Ω; radio and television or any such transmission sites should have an earth resistance below 5 Ω. The stipulated resistance for transmission and distribution lines, radio and television, and transmission sites *shall* be achieved and maintained at all times. It is recommended that all joints are cad welded to eliminate connection becoming loose. The reason for this is simple; utility poles and towers are in most cases above normal habitats and are the first point of contact for lightning. Therefore, if the low resistivity resistance is not achieved at each of these points of contact or within the distribution network, lightning damage will be inevitable and can be tremendously costly and sometimes fatal.

6.4 Supplementary Grounding

Supplementary grounding is the addition of one or more earth electrodes or grounding systems which gives support to the main grounding system. It is also used for providing additional protection for outdoor equipment. Supplementary grounding is optional in some situations but is mandatory in several other situations. Its use is dependent on the type of soil (rocky, sandy, stony, etc.), the type of environmental conditions (lightning-prone or dry), and the type of installation (residential, commercial, or Industrial). Supplementary grounding is essential in installations such as power stations and substations or any other such type of installation where there may be high potential appearing on the main earth conductor.

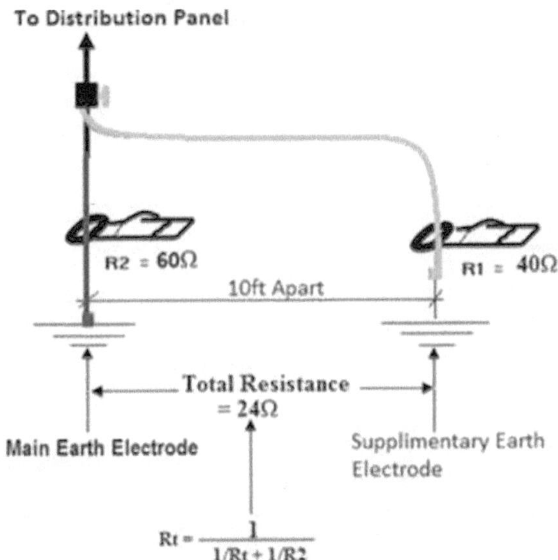

Figure 6.5: *Achieving effective ground resistance using supplementary grounding*

6.4.1 Examples of Supplementary Grounding System

Machines or sub-circuits and panels which are located long distances from the MDP may use supplementary grounding electrodes. These electrodes, when connected to these machines or panels, serve to reduce the effect of HV induced by lightning. The additional protection prevents high potential from reaching equipment or sensitive appliances by diverting such potentials to the general mass of earth. In order to ensure optimum safety all outdoor equipment or machinery must be linked to a supplementary electrode or a supplementary grounding system as shown in Figure 6.6.

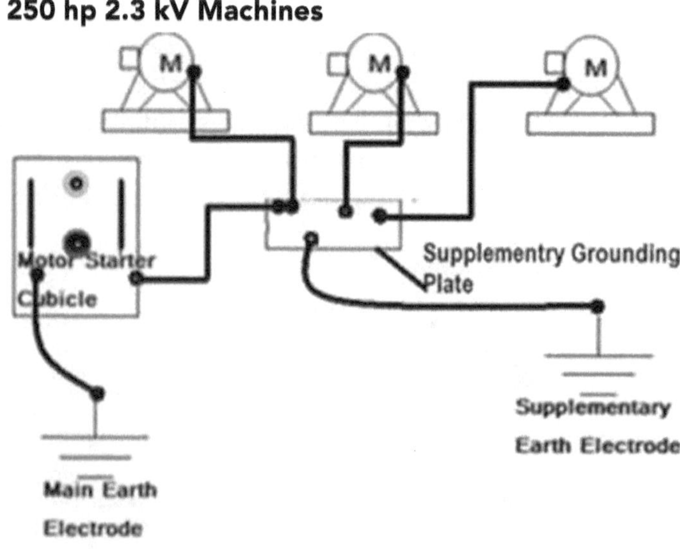

Figure 6.6: *Supplementary earth plate and electrode*

Another example of supplementary grounding is grounding provided for steel towers and structures. Bare copper is cad welded to the steel structure, encircling the tower or building. Figure 6.7 demonstrates a typical arrangement. Other examples of supplementary grounding will be demonstrated in the next chapter.

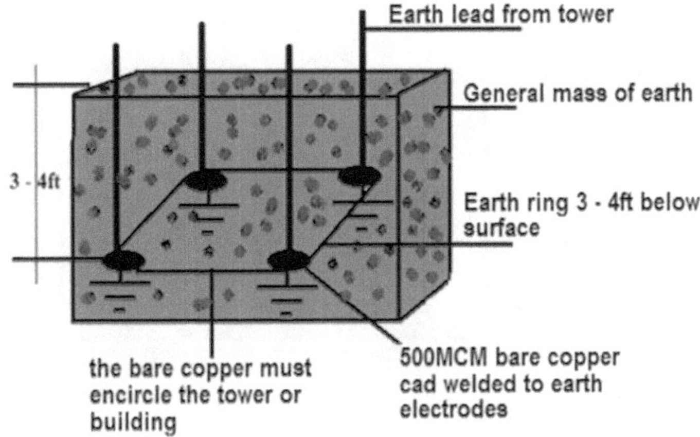

Figure 6.7: *Grounding steel towers or structures*

The resistance between the general mass of earth and the earth electrode is sometimes so low that 0 Ω is reflected on the measuring instrument. In other cases, it is unacceptably high, that is, above 200 Ω. An extremely good connection to the general mass of earth will reflect a minimum resistance reading of 0.3 Ω. However, readings of 0.3 Ω or below should be thoroughly investigated to ensure that there are no irregularities with the grounding configuration.

6.5 Sizing of the Main Earth Conductor for an Installation

The main earth or ground conductor is the conductor which connects the entire installation to the general mass of earth. Therefore, this conductor size should be chosen based on the maximum short-circuit current available on the installation, or, as a rule of thumb, as shown in Figure 6.8, the main earth conductor should be minimum "half the size of the largest phase conductor of the circuits" on the installation. In addition, fault which travels through the main earth conductor during a short circuit is only a small portion of the maximum fault current available on the installation.

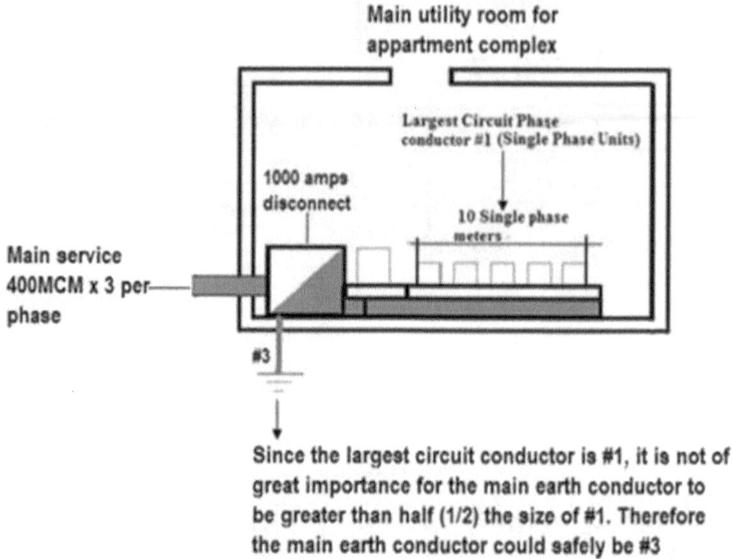

Figure 6.8: *Determining the size of the main earth conductor for an apartment complex*

A high-impedance grounding system will result in a volt drop between the electrical panel and the general mass of earth, thus causing significant current to flow to the earth. Figure 6.9 shows the result of short-circuit current to the earth electrode versus short-circuit current to the neutral conductor.

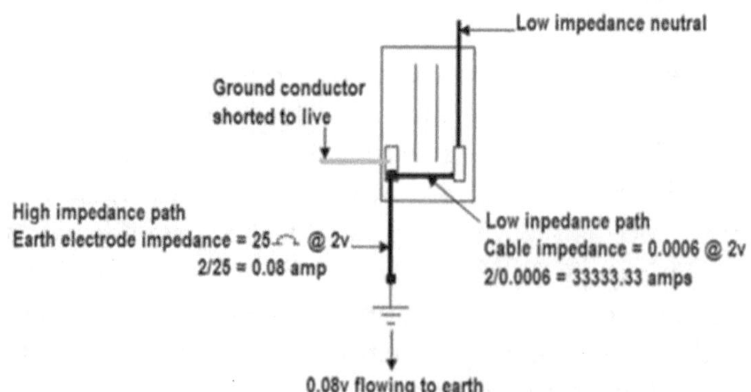

Figure 6.9: *Short-circuit current for earth electrode vs. short-circuit current for neutral conductor*

IEE wiring regulation, sixteenth edition, gives guidance on the methods of choosing the main earth conductor for an installation:

Acknowledgements to the **BS 7671, "Requirements for Electrical Installations", and the BSI and IET, with all rights reserved to the BSI and IET. The information below has been "Reproduced by permission of IET".**

Section 543-01-02: Where a protective conductor is common to several circuits, the cross-sectional area of the conductor shall be;

- Calculated in accordance with Regulation 543-01-03 for the most of the value of fault current and the operating time encountered in each of the various circuits, or
- Selected in accordance with Regulation 543-01-04 so as to correspond to the cross-sectional area of the largest phase conductor of the circuits.

Section 543-01-03: The cross-sectional area, where calculated, shall not be less than the value determined by the following formula or shall be obtained by reference to BS 7454

$$S = \frac{\sqrt{I^2 t}}{K}$$

Where:

S is the nominal cross sectional area of the conductor in mm^2

I is the value in amperes (r.m.s. for a.c.) of fault current for a fault of negligible impedance, which flows through to the protective device.

t is the operating time of the disconnecting device in seconds corresponding to the fault current in amperes

K is the factor taking account of the resistivity temperature coefficient and heat capacity of the conductor material and the appropriate initial and final temperatures

6.6 Earth Conductor Specifications

It must be reiterated that at no time will the main earth conductor of an installation experience particularly high current and voltage during a short circuit. The TNCS earthing configuration ensures the earth conductor is not compromised as a result of high currents. Large cables (above the sizes prescribed in Table 6.1) will not increase the safety of an installation. However, it is imperative that specifications of this table be adhered to when designing earthing systems. Additionally, guidelines provided by the authority having jurisdiction or the chief electrical inspector shall be followed.

The r.m.s. values used in Table 6.1 are maximum short-circuit current values. The time values are the time protective devices take to operate, and *K* is a value taken from the IEE Regulation Table 54B.

Table 6.1 Specifications for main earth conductors

kVA	Amp	RMS Value kA @ 208/120V Y	$S = \frac{\sqrt{Isq.t}}{K}$ = Amp	Main Earth Conductor	
5 - 75	up to 200	10	42	#8	4mm²
75 - 150	200 - 400	22	97	#6	10mm²
150 - 300	400 - 800	42	185	1/0	95mm²
300 - 500	800 - 1350	65	287	4/0	50mm²
500 - 750	1350 - 2050	65	287	4/0	95mm²
1000	2050 - 2700	85	375	300mcm	150mm²

kVA	Amp	RMS value kA @ 480/277V -Y	$S = \frac{\sqrt{Isq.t}}{K}$ = Amp	Main Earth Conductor	
5 - 75	up 90	14	62	#6	10mm²
75 - 150	90 - 180	14	62	#6	10mm²
150 - 300	180 - 360	25	110	#2	25mm²
300 - 500	360 - 600	30	133	#2	25mm²
500 - 750	600 - 900	30	133	#2	25mm²
1000	900 - 1200	50	221	2/0	50mm²
1500	1200 - 1800	50	221	2/0	50mm²
2000	1800 - 2400	50	221	2/0	50mm²
2500	2400 - 3200	55	243	3/0	70mm²

According to Article 250.66 of the NEC, cable sizes need not be greater than those provided in Table 6.1. Additionally, the main earth conductor for any installation may be derived by using a percentage of the designed current as shown in Table 6.2.

The main circuit breaker of any installation is designed to operate manually for maintenance or emergency purposes. In addition, the circuit breaker is also designed to operate automatically if the following conditions are satisfied:

1. There is a short circuit between the phase terminals on the panel bus bars
2. There is a short circuit between live bus and ground terminals
3. There is a short circuit between the live bus bars and the neutral terminals

Table 6.2: Calculating main earth conductor sizes using percentages

Designed Current	Percentage (%) Designed Current, Matching Main Earth Conductor	Short-circuit Disconnecting Time	Earthing Arrangements and Voltage
Up to 200	24 (%)	1.5-5 sec	TT-120/240 V
200-400	24 (%)	0.8 sec	TN-120
400-1800	16 (%)	0.4 sec	TN-120/240 V
200-1500	22 (%)	0.2 sec	TN-220/415 V to 277/480 V
1500-3000	15 (%)	0.2 sec	TN-220/415 V to 277/480 V

Therefore, based on the earth and neutral combined (TN) earthing arrangement described in Table 6.2, the fault current acting on the main earth conductor during a short circuit will not be greater than the calculated values in Table 6.1 or the percentage values which may be derived from Table 6.2.

6.7 Conclusion

Grounding systems are designed to protect and provide a path for unwanted currents and voltages to be dissipated to the general mass of earth. Grounding also eliminates noise or interferences from electrical systems.

Grounding is crucial for all electrical installations. It is essential to achieve good connectivity between the general mass of earth and the earth electrode in order to provide optimum protection from electrical fault currents. When grounding electrical installations, it is essential that the lowest possible resistance is attained. This should not exceed an average of 120 Ω for domestic installations. Equipment sensitivity and soil conditions must be assessed to ascertain whether additional electrodes may be needed for reducing earth resistance and hence faster removal of fault current from the installation. If resistance measurements below 25 Ω cannot be achieved with one electrode, additional electrodes must be driven to support the main electrode. In situations where the soil quality is poor, specially designed chemical earth pit may be constructed to achieve low resistances

Supplementary grounding is essential in providing grounding in areas where adequate grounding cannot be achieved with one earth electrode. This grounding is also essential in the elimination of touch and step potentials.

6.8 Test Your Knowledge

1. State three reasons for employing a good grounding system.
2. What is the minimum and maximum acceptable earth reading for installation based on your local electrical code?
3. When does it become necessary to install a second earth electrode for an installation?
4. What is meant by the phrase "general mass of earth?"
5. Why is it critical to maintain earth readings below 200 ohms (Ω)?
6. The dissipation of voltage and current to the general mass of earth is dependent on three factors. State these factors.
7. Why is it important to eliminate noise on electrical systems?
8. What is the main aim of an earth electrode in an installation?
9. Write and explain the formula used to calculate the size of the main earth conductor of an installation.
10. What is the minimum earth conductor size to be used in an installation?
11. What is meant by "half the size of the largest circuits phase conductor?"
12. Outline and explain the principle of operation for three instruments used to measure soil resistivity.
13. What is the best method of connection for reducing electrical noise?
14. List four devices which influence noise.
15. An installation has a short-circuit current of 6000 A. Using the word *minimum* or *maximum*, how much of this current would be found on the main earth conductor. Give reasons for your answer.

Bibliography

i. Avallone, E. A. and Baumeister, III T. 1986. Marks' *Standard Handbook for Mechanical Engineers*. The Ninth ed. Page 6-62

ii. IEEE Std 142-1991. IEE Recommended Practices for Grounding of Industrial and Commercial Power Systems. Retrieved on June 14, 2012, from https://www.academia.edu/4088942/EEB._GREEN_BOOK_IEEE_Recommended_Practice_for_Grounding_of_Ind Johnson, Nigel, 2006., Earthing Manual. Retrieved April 29, 2014 from: http://www.scribd.com/doc/60854555/E3-Earthing-Manual-Soil-Resistivity

iii. Piantini, A., n.d. Lighting Protection of Overhead Power Distribution Lines. Retrieved April 29, 2014 from: http://www.ebookbrowse.com/invited-lecture-4-pdf-d41833750

iv. Lyncole XIT Grounding. Soil Resistivity, Testing Four Point Wenner Method. Retrieved on June 14, 2012, from http://www.lyncole.com/uploads/Lyncole_Ground_Test_Methods.pdf

v. Metal Gems. 2008. Copper Earthing Systems. Retrieved on June 14, 2012, from www.mehta-group.com/copper-earthing-systems.html

vi. Myers, J. R. and Arthur, C. August 1994. Conditions Contributing to Underground Copper Corrosion. *American Water Works Association Journal*. Retrieved on June 14, 2011, from www.copper.org/resources/properties/protection/underground.html

vii. National Fire Protection Association. 2008. National Electrical Code National Fire Protection Association. Seventeenth ed. Page 44

viii. Nennemann, G. R. and Mantel, S. 1995, Soil Brief Jamaica 1. Retrieved on October 14, 2010, from http://www.isric.nl/ISRIC/Webdocs/Docs/soilbrief_Jamaica01.pdf

ix. Mike Holt Enterprises. National Electrical Code Internet Connection. Posted on December 30, 1999. Opening Circuit Overcurrent Protection Device to Clear Line-to-ground Fault. Retrieved from http://

www.mikeholt.com/mojonewsarchive/All-HTML/HTML/GFCI-Receptacles-Without-Ground~19991230.php

x. Requirements for Electricians. IEE Wiring Regulations. Sixteenth ed. Page 27-31, Wiring Systems

xi. Russell, M. J. 2000. The Impact of Mains Impedance on Power Quality. Retrieved on June 14, 2012, from http://www.powerlines.com/pq2kdoc.pdf

xii. Schneider Electric. June 2003. *HV Training Manual.* Chapter 7

xiii. TECHTIP 60201. How to Reduce System Noise, n.d. Retrieved from http://www.mccdaq.com/pdfs/techtip/techtip_60201.pdf

xiv. US Department of tTransportation fFederal hHighway aAdministration. Corrosion/Ddegradation of SSoil RReinforcements for MMechanically SStabilized EEarth WWalls and RReinforced SSoil SSlopes. Publication No.FHWA-NHI-00-044. Retrieved on June 14th, 2012, from: http://www.scribd.com/doc/87701092/7/IDENTIFICATION-OF-CORROSIVE-ENVIRONMENTS

CHAPTER 7

Improper Grounding Hazards and Special Grounding

7.0 Introduction

Improper grounding hazards refer to conditions where damage to property and life is threatened as a result of improper grounding. In these circumstances, the earthing facility is inadequate for dissipating leakage current to the general mass of earth. This condition can prove dangerous and, in some cases, fatal. Some of the conditions resulting from improper grounding are discussed in this chapter.

7.1 Touch Potential

Touch potentials are voltages which appear on grounded or ungrounded metal frames due to a fault. These voltages may be intermittent or/and may be sustained in the affected area over a period of time. Figures 7.1 and 7.2 depict *touch potential,* which could be fatal. Touch potential can be eliminated by securely bonding all metal parts (exposed conductive parts) to a common grounded system.

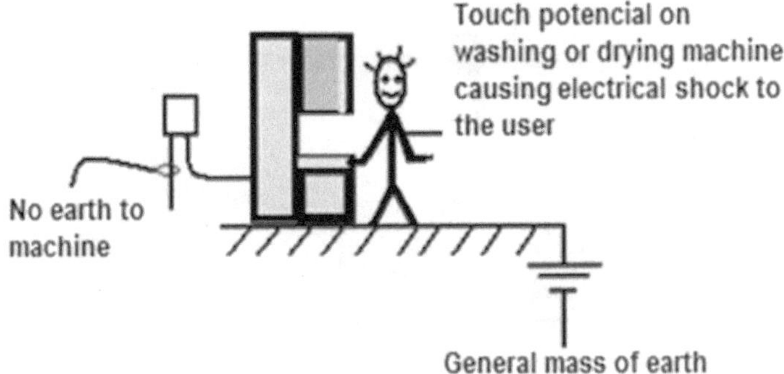

Figure 7.1: *Touch potential resulting in electrical shock*

Figure 7.2: *Machine connected to earth diverting touch potential to earth*

7.2 Step Potential

Step potential is a change in voltage per foot step or per meter on the surface of the general mass of earth. This potential is due to the variation of the soil resistivity between the main earth electrode of electrodes and the supply transformer.

Since the resistance of the human body has an average resistance between 300 Ω to 1500 Ω, the resistance of the body while in contact with the surface of the general mass of earth per meter may be lower than the resistivity of the general mass of earth. Therefore, the body, if not appropriately insulated, will act as the easiest path for a fault current, causing injury or fatalities. Figure 7.3 depicts relatively HV between walking steps, which could prove fatal in the vicinity of a HV transformer which is not properly grounded.

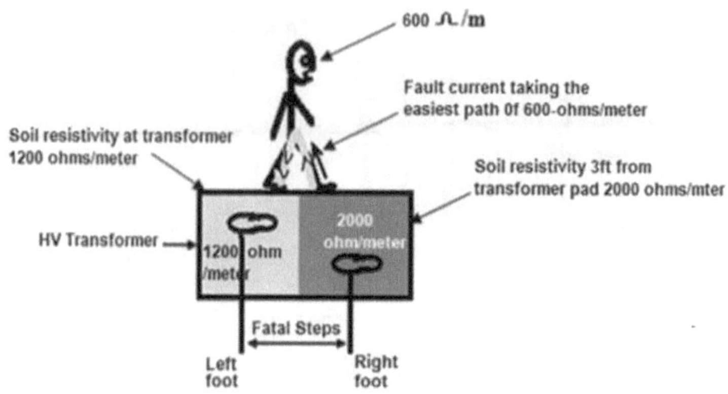

Figure 7.3: *Dangerous step potential in vicinity of poorly grounded HV equipment*

7.3 Eliminating Step Potential

Step potential can be eliminated by creating a grounding grid or grounding mat. A grounding mat shown in Figure 7.4 is a supplementary ground system which suppresses and equalizes high potential acting on the general mass of earth, thereby eliminating dangerous deferential step potential across the surface of the general mass of earth.

7.3.1 Grounding Mat

A grounding mat is a number of copper conductor horizontally laid and clamped or welded together, forming a number of square grids. This network is also connected to a number of earth rods equally spaced at 1 m apart as shown in Figure 7.4.

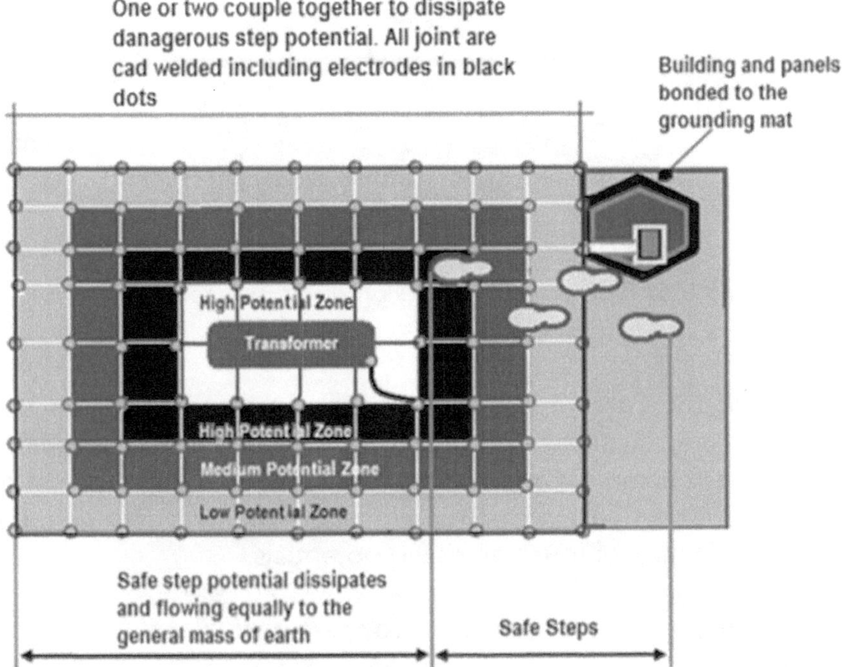

Figure 7.4: *Typical grounding mat to eliminate step potential and touch voltage*

7.3.2 Low Resistivity Ground Mat

In instances where the soil quality is poor, grounding mats will prove ineffective in providing a low resistance connection to the general mass of earth. In such instances, specially designed grounding facility is necessary to establish a low resistance connection. Figure 7.5 shows a grounding mat embedded in a low resistivity compact soil pit, sometimes referred to as a chemical earth pit. The low resistivity compact soil consists of a combination of salt, coal, clay, and other resistance-lowering materials.

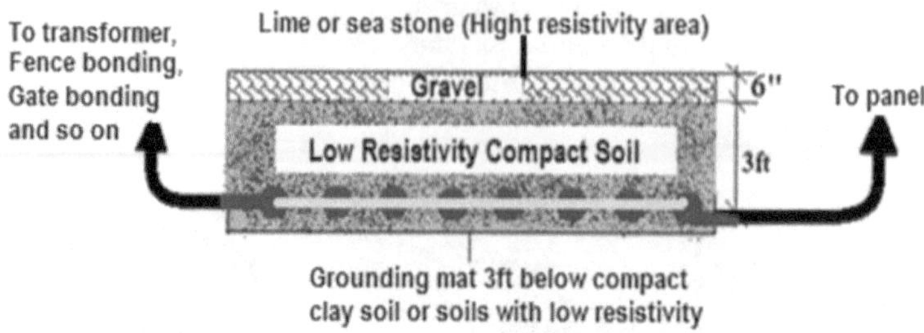

Figure 7.5: *Typical low resistivity grounding mat arrangement (chemical pit)*

Grounding mats are not standard or a requirement for domestic or commercial installations. Supplementary grounding is essential in providing grounding in areas where adequate grounding cannot be achieved with one earth electrode. This grounding is also essential in the elimination of touch and step potentials.

7.3.3 Bonding Arrangements for Ground Mats

Two methods used for providing bonding for grounding mats throughout the world are grounding plates and rebar or structural foot grounding connectivity as shown in Figure 7.6.

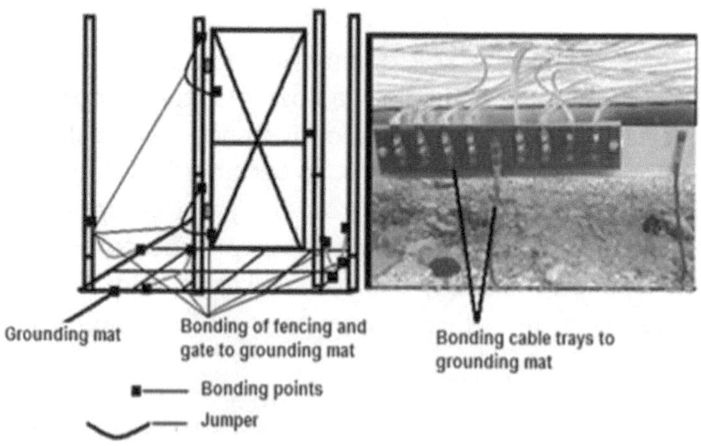

Figure 7.6: *Typical bonding arrangement for grounding mats*

In instances where rebars are used as a part of the grounding system, one earth rod must be planted at each corner of the foundation footing of single-story buildings and at each corner and columns for two-story structures as shown in Figure 7.7. The earthing arrangement shown in Figure 7.7 will reduce the effect of touch potential and lightning. This type of arrangement is recommended for apartment buildings, convention centers, hotels, or any such areas where a large number of people may be gathered or for building where sensitive equipment is housed. The earthing arrangements shown in Figures 7.7 and 7.8 are cheaper, not easily affected by soil erosion, and are quite easy to install using regular electrical tools. Grounding plates should be no less than 2 $ft^2 \times$ ¼″ in dimension. This recommendation is supported by all international electrical codes.

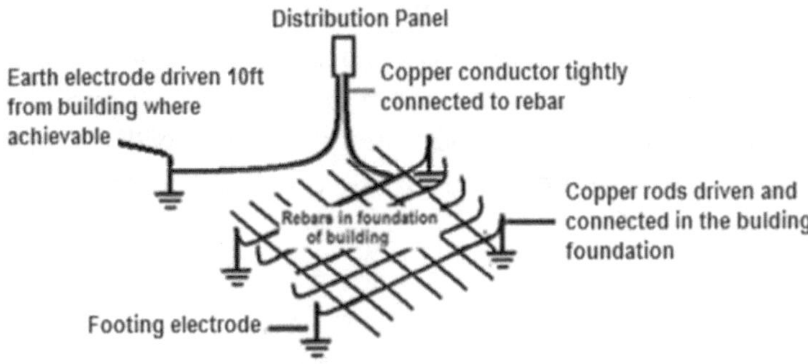

Figure 7.7: *Using rebar to supplement earthing*

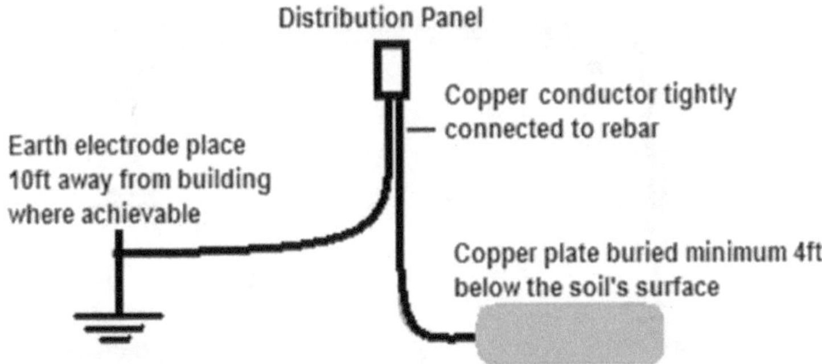

Figure 7.8: *Using copper plate to supplement grounding*

7.3.4 Special Bonding

Protective conductors may become damaged as a result of mechanical stress on the cables, the location of cables, or the external ambient air condition of the cables. Cables running on cable racks or cable trays are more likely to be exposed to these conditions and may put a grounding conductor at risk of becoming damaged. Conditions affecting cables include, but are not limited to, the following:

1. Corrosion due to poor air quality, e.g., in a bauxite environment (acidic air)
2. Corrosion due to direct contact with other corrosive metals

Cable racks and cable trays must be bonded to the equipment panel and equipment which they serve. Using supplementary electrodes, bonding plate, or plates will provide increased grounding protection to the main grounding electrode of the installation. This type of grounding assembly will provide easy monitoring and maintenance of all grounding terminations as shown in Figure 7.9.

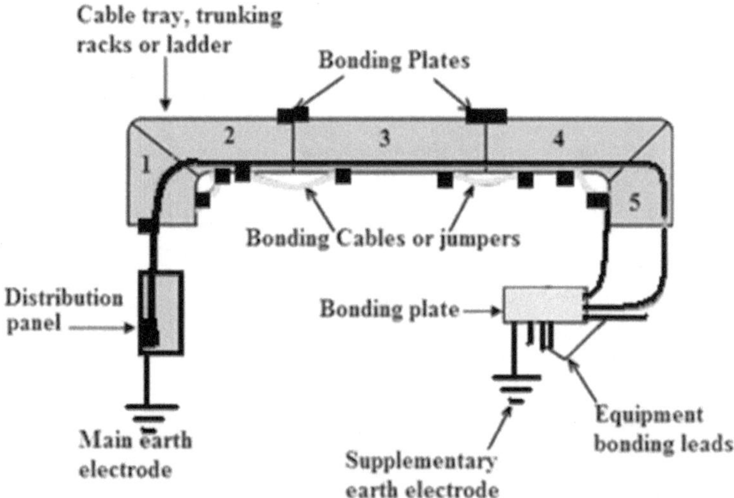

Figure 7.9: *Typical bonding of cable trays, trucking, racks, and ladder*

7.4 Resistance Grounding

Resistance grounding is a grounding system which capacitively couples each phase of a three-phase system to the ground. In addition to this, the star point of the supply transformer is connected to the general mass of earth through a resistor. Thus, the system is referred to as an ungrounded system and is used in many processing industries where processing equipment cannot be shut down because of faults which are manageable.

Ungrounded systems are defined as systems without a direct grounding connection to the general mass of earth. These systems are connected through electronic components and measuring devices to safely monitor and re-circulate safe fault current through the protected system. The load condition of an ungrounded system is reasonably balanced; therefore, the neutral points of these systems are usually close to ground potential i.e. 0 V. To further clarify, there is always a capacitive coupling between the phase conductors and a dedicated solid grounding system; hence, an ungrounded system is usually referred to as a *capacitively* coupled grounding system.

Ground fault units or resistance grounding units are available in two types: high resistance and low resistance. A low resistance unit operates at fault levels between 200 A and 1200 A, while a high resistance unit operates at current levels between 5 A and 10 A. In addition, low resistance systems operate at LSs and HVs, and high resistance grounding operates at voltages below 600 V with limited application to motors with voltages at 2.4 kV and 4.6 kV. Figure 7.10 shows a typical ungrounded grounded system.

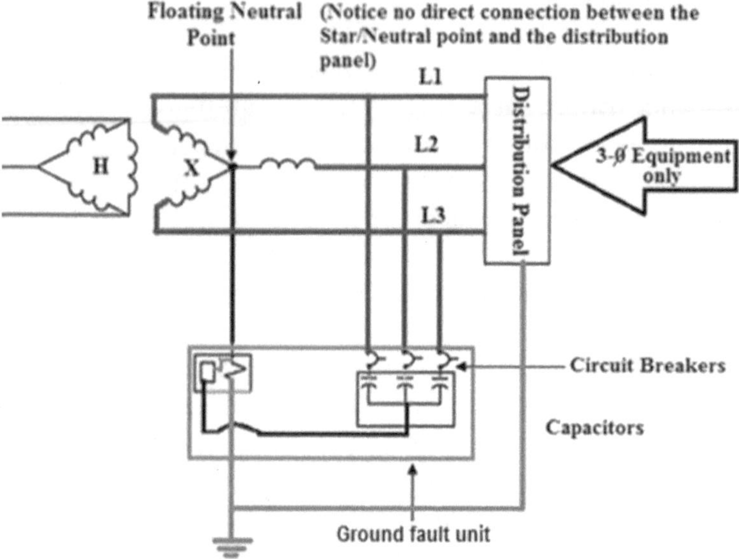

Figure 7.10: *Typical ungrounded configuration*

7.5 Ungrounded Faults

Ungrounded faults are faults which occur in ungrounded distribution systems. An ungrounded system has no direct connection between the star and neutral point of the supplying transformer and the supplied equipment. The same is true for *delta/delta* ungrounded system. The low resistance grounding system is a very effective ungrounded system and operates similar to a solid grounding system. This system, due to its unique current-monitoring and tripping ability, does the following:

1. Significantly reduces damage to connected equipment
2. Prevents further catastrophic damage from occurring
3. Provides personal safety
4. Keeps faults within the supplying circuits
5. Prevents overheating and torsional oscillation

High resistance grounding is an ungrounded system which can be installed easily on any active grounding system, without incurring cost for additional protective devices. This device, due to its ground fault current-limiting characteristics, allows a fault to be sustained and recirculate on a system during production until the ground fault can be located.

The disadvantages of ungrounded systems are as follows:

1. After an initial ground fault, the voltage of the two other phases rises to the line-to-line voltage. For example, if the voltage of an ungrounded system is 480 V, the total line voltage will be circulating through two phases instead of circulating through three phases. This results in an increase in voltage stress of 173% on the insulation of the system and the other two unfaulted phases.
2. When a ground fault occurs, the star point cannot be used as a neutral point for load connections such as single-phase lighting. The neutral point of the system rises to line voltage, which is significantly above ground potential.
3. If there is a ground fault on another phase before the first fault is corrected, a line-to-line fault is created. This could be catastrophic.

7.6 Zigzag Transformer Grounding

A zigzag transformer is a dry-type air cool auto transformer with 6 windings. Each pair of windings is connected in reverse to each other and has the same number of turns. This type of configuration forces the three-phase currents to become in phase with each other; this is referred to as zero sequence current. This winding operation causes a fault current to return into the system, thus limiting the fault current and the damage it may cause.

This type of transformer has multiple industrial purposes. One of its most common uses is for grounding an ungrounded system. There are other types of auto transformers; however, the most frequently used grounding transformer is the zigzag transformer. Zigzag transformers as shown by the representation in Figure 7.11 do not have a secondary winding. These devices are fitted with circuit breakers or fuses and are connected to the line side distribution systems, preferably close to the distribution transformer.

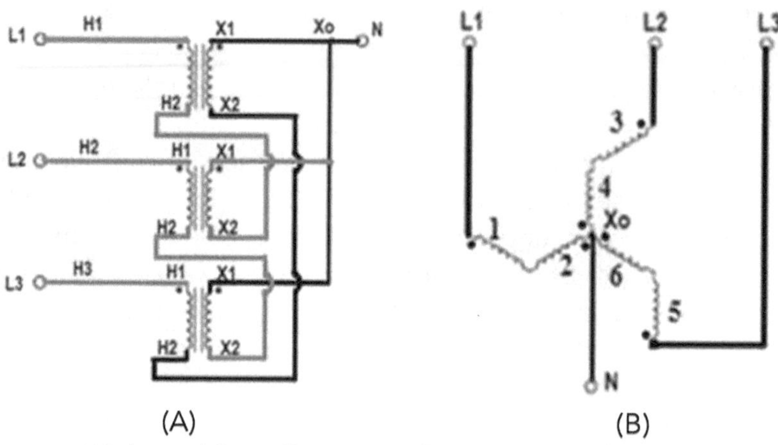

(A) (B)

(Adapted from Section 6: System Grounding, n.d.)

Figure 7.11: *(A) Typical zigzag winding configuration*
(B) Typical circuit representation

Similar to the low and high resistance grounding system, the terminals of the zigzag transformer are connected to the load side, line 1, line 2, and line 3 of the distribution transformer as shown in Figure 7.12 where the star or neutral point of the transformer is connected through a resistor and then to the general mass of earth.

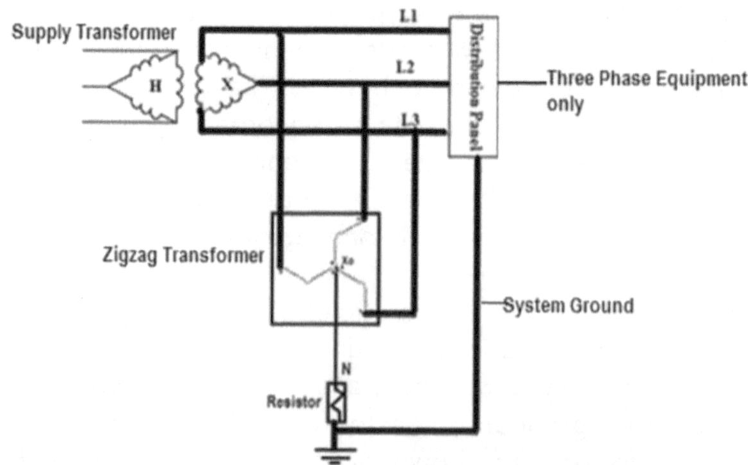

(Adapted from Section 6: System Grounding, n.d.)

Figure 7.12: *Typical zigzag transformer connection*

7.7 Disconnecting Grounding Cables

Grounding cables must not be taken for granted. Many electricians and technicians are unaware of the significance of this resource and the dangers associated with it especially in industrial plants and commercial enterprises that uses significant amount of electrical power.

Disconnecting grounding cables while power is supplied to facilities could prove fatal if not done properly. It is therefore advised that one should "Never attempt to disconnect an earthing conductor, from its point of termination while troubleshooting a fault or testing, and while a plant is in full operation." Figure 7.13 depicts what could take place if a grounding cable is improperly disconnected.

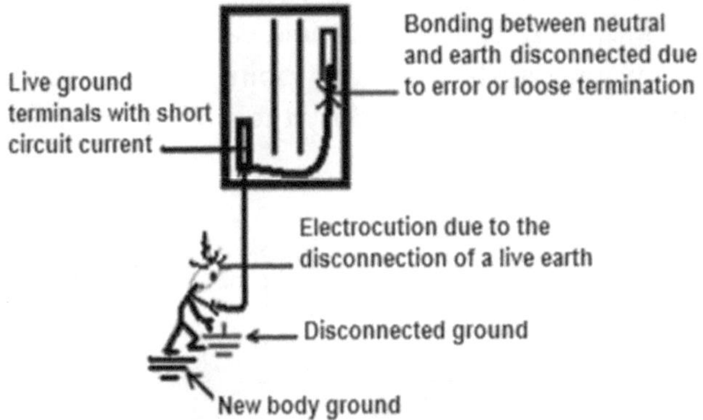

Figure 7.13: *Electrocution due to disconnecting live earth conductors*

Grounding cables should be disconnected only after the power source to an installation has been isolated and reconnected prior to restoring power to the installation.

7.8 Creating a Solid Grounding from an Ungrounded System

Before connecting single-phase loads to panels in an industrial environment, care must be taken to identify the appropriate equipment or appliance panels. In most instances when other equipment or appliances are coupled as is shown in Figure 7.14, the unit may give a ground fault error due to the increase of the neutral or ground current, resulting in system failure. In addition, appliances connected to this panel are likely to be damaged from HV should a fault occur.

Problems associated with ungrounded systems can be resolved by retrofitting a delta/star isolation transformer or a *delta/star* step-down transformer to supply sub-panels or lighting panels as shown in Figure 7.15. These panels can then be connected to a new solid ground system. At this point, zero sequence current will not affect the loads supplied by the ungrounded system. See Figure 7.15.

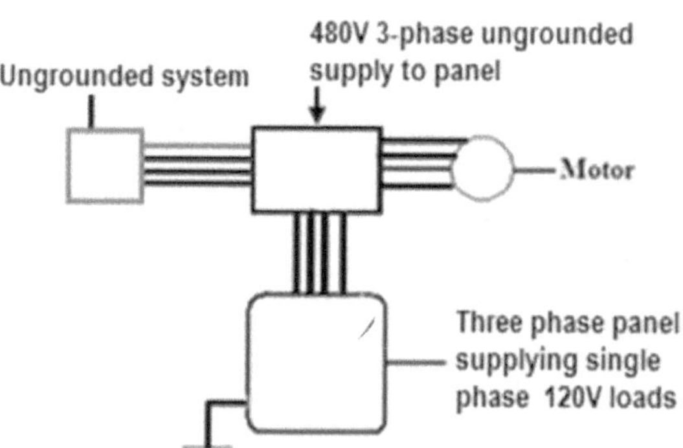

Figure 7.14: *Supply sub-panel from an isolation transformer*

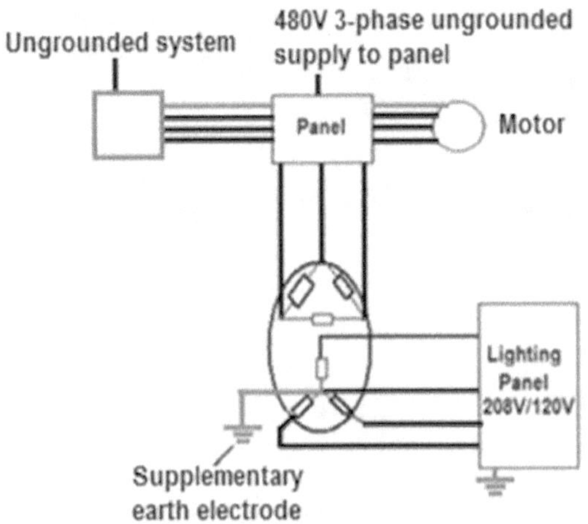

Figure 7.15: *Connecting a lighting panel to an ungrounded system*

7.9 Parallel Path

"A parallel path is a loop which allows dangerous current to circulate in an installation. This current can lead to serious injuries and damage to equipment and appliances." In literal terms, a parallel path is the bonding of the service entrances neutral and earthing conductors at two points on the installation (the meter center and the MDP). If a parallel path condition exists in an installation and there is an open neutral on the utility transformer or at any point between the transformer and the distribution panel, dangerous current will be in circulation while 120 V appliances are in use.

Current circulates wherever a loop exists as shown in Figure 7.16. The circulation continues until all potential dissipates to the general mass of earth or until the voltage reaches 0 V. The effect of a parallel path can be felt on exposed conductive parts or accessories such as metal boxes, electric stoves, washing machines, refrigerators, and other electrical household equipment. The most common effect of parallel path is intermittent electrical shock where a fault to ground is present.

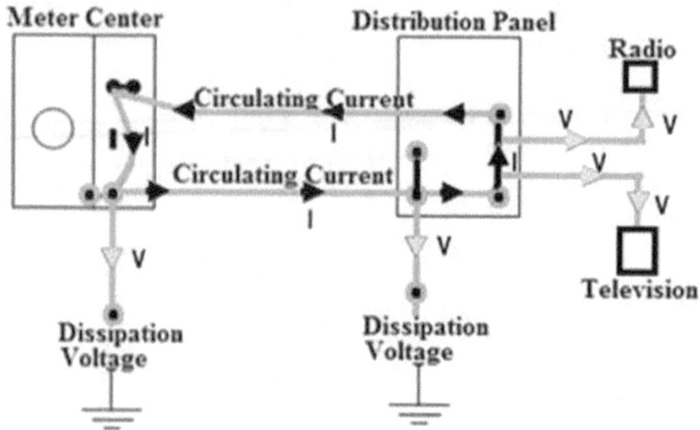

Figure 7.16: *Shows dangerous current (I) in circulation and dissipating voltage (V)*

When an electrical installation does not have a parallel path, it does not mean that there is fault current in the installation; however, fault current and voltage present on an installation have an easy path for dissipation to the general mass of earth. Where parallel path does not exist, the effect of touch voltage and current is significantly reduced and further reduces the possibility of damage to equipment during lightning, storm, and other similar faults.

Parallel path in any installation acts as an arc ring; this means that switching off appliances without unplugging from the outlets will not protect appliances from lightning and transients. Figure 7.16 shows the behavior of voltage and current in a parallel path during lightning and transients, and the effect it has on appliances. Importantly, a parallel path provides multiple avenues for fault currents and allows fault current to bypass protective devices on the systems.

According to the NEC, "A careful engineering study must be made to ensure that fault currents do not take parallel path to the supply system, thereby bypassing the ground fault detection device."

Bonding should be done at the main disconnect or at the first point of service entrance for multiple dwelling or apartments and complexes with multiple meters. This type of bonding should be a rule of thumb to reduce the complexity of troubleshooting the grounding system.

Switching off appliances without unplugging them from the outlet will not protect appliances from lightning and transients. However, the chance of an appliance being damaged from transients is significantly reduced. As observed in Figures 7.17 and 7.18, the current and voltage dissipating to the general mass of earth does not interact with the neutral conductor of the appliance.

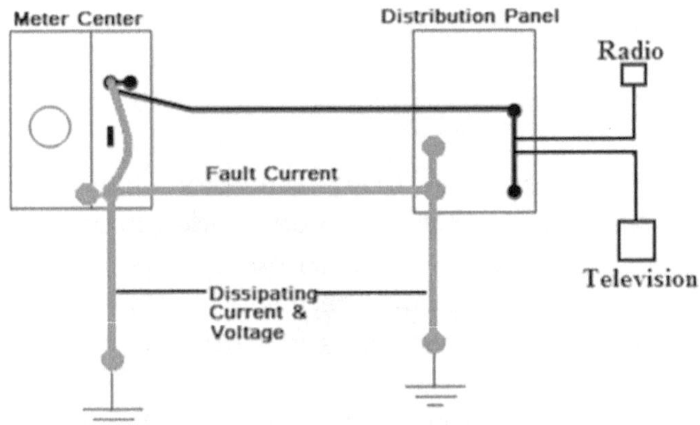

Figure 7.17: *Transient current and voltage dissipating safely to the general mass of earth*

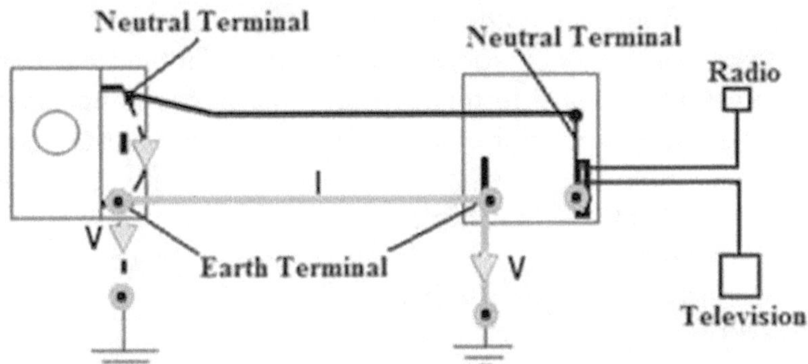

Figure 7.18: *Installation working under normal condition with no parallel path. Note that the jumper is removed*

7.10.1 Key Points to Remember about Parallel Path

1. High current will not dissipate quickly from the effects of transients and lightning, thus causing back-feed through the neutral conductor. As a result of the back-feed, the neutral conductor could cause damage to 120 V electrical appliances.
2. It is highly possible for potential coils of kWh (kilo watt hour) meters to become damaged.
3. The possibility for trapped current initiating fires which may result in wooden structures becoming engulfed in flames and causing extensive damage to property.

7.10 Bare Copper Grounding

Bare copper used as an earthing conductor for sub-circuits and other earthing requirements should be avoided except for outdoor direct burial. This practice is an extremely unreliable method of wiring. Except for underground earthing, configuration, or burial, the use of bare copper conductors for normal circuit wiring is not considered a safe working practice. Bare copper with fault current will allow that fault current to circulate to other circuits, which initially have no faults.

If there is a phase-to-ground fault on a circuit which passes through a conduit with other circuits with the ground conductor being bare copper, it is highly possible that this fault current will affect other circuits inside this conduit. In this case, the effect of this fault is not limited to the point of origin of the fault but several other areas within the installation. In this event, difficulties may be experienced when troubleshooting since the fault current will be present on several bare copper conductors, which may require a total rewiring to eliminate the fault.

7.11 Conclusion

There are several hazards which may result from improper grounding. Hazards such as touch and step potentials are two significant hazards which must be eliminated at all cost. Touch potentials are voltages which appear on grounded or ungrounded metal frames due to a fault. Supplementary grounding assists in the elimination of touch potentials. There are other grounding systems known as special grounding systems which are used in large production plant and refineries to reduce or absorb ground fault currents. This type of grounding allows an equipment or system to operate without shutting down instantaneously.

7.12 Test Your Knowledge

1. What are the reasons given for grounding mats which are not included in the electrical code?
2. Why is it important to install supplementary earth electrodes?
3. Define what is meant by parallel path.
4. Describe the behavior of fault current through a parallel path.
5. List the dangers of an improper earthing arrangement.
6. Sketch a diagram of an earthing arrangement which will allow current and voltage to dissipate to earth, and explain the behavior of fault current through this arrangement.
7. List two alternate methods used to reduce earth electrode resistance.
8. Draw and label a good earthing arrangement.
9. Why is the use of bare copper considered a bad trade practice?
10. During a lightning storm, what is the safest way of protecting appliances? Explain.
11. What is the best method of connecting a lighting panel to a resistance grounding system?
12. What is the best wiring method of eliminating touch voltage?
13. What is the suggested method of grounding used to monitor grounding current on critical equipment or machinery?
14. What is the safest method of disconnecting an earth conductor from a live system?

Bibliography

i. Bill, B. n.d. Square D Engineering Services. Section 6: System Grounding. Retrieved on August 5, 2012, from

ii. Beltz, R. Cutler-Hammer, 3900 Kennesaw 75 Pkwy. Atlanta, GA 30144

iii. Johnson, G, n.d., Basler Electric. Schroeder, Dominion VA Power. Dalke Gerald. Power Systems Relay Service. A Review of System Grounding Methods and Zero Sequence Current Sources. n.d. Retrieved on August 5, 2012, from http://www.blaster.com/downloads/ReviewOFSysGrounding.pdf

iv. Peacock, I. n.d. Cutler-Hammer, RR3 Box 910F, Winthrop, ME 04364. Vilcheck, W. Cutler-Hammer, 130 Commonwealth Drive, Warrendale, PA *15086*. Application Consideration for High Resistance Grounding Retrofit in Pulp and Paper Mills. Retrieved on August 5, 2012, from http://www.tappi.org/Downloads/unsorted/UNTITLED-pcei0133pdf.aspx

CHAPTER 8

Earthing Configurations

8.0 Introduction

Earthing configurations are essential for the purposes of grounding electrical installations to reduce the risk of electric shock. Significantly, earthing configurations are also crucial for protecting domestic, commercial, and industrial appliances and equipment from the destructive energy that may be created from a short circuit at any point of the installation. Earthing configurations dictate the path that fault current or transients take to dissipate to earth. The general earthing configurations used on electrical installations in some countries such as those of the Caribbean, Europe, and North America are the focus of this chapter.

8.1 TNCS and TT Systems

In the TNCS system T represents a single earth electrode placed vertically to the general mass of earth, forming a *T* before it is driven into the earth N represents "neutral", C represents "combined, and S represents "single conductor".

The TNCS system uses an earth configuration in which the neutral terminal and the earthing terminal are combined in a single conductor. This means that an earthed neutral conductor is supplied by the electricity supplier to the consumer without a separate earth conductor to the consumer's meter facility. At the consumer's meter facility, an earth electrode is planted, and, at this point, a separate earthing conductor is created by linking the neutral and the earthing terminal. Since a main earth terminal is created, an earth conductor along with the supplied neutral is taken to the main distribution board of the installation as shown in Figure 8.1. The TNCS system does not require earth electrodes at the consumer's installation.

This system provides additional earthing and in some cases combined with the TNCS system. The combined configuration is called the "multiple earthing system". Each multiple earthing system is an earthing configuration which provides multiple low-impedance pathways for fault current to the general mass of earth. Figures 8.2 and 8.3 provide diagrams of these configurations.

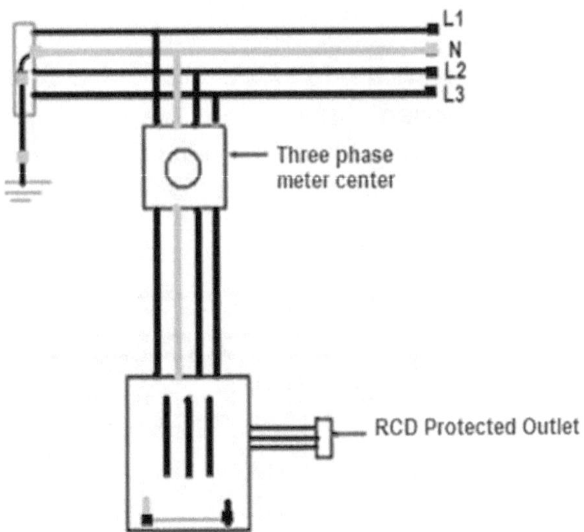

Figure 8.1: *TNCS—150 kVA, 208 V, 3 Ø transformer supplied power with no earth electrode at the consumer's installation*

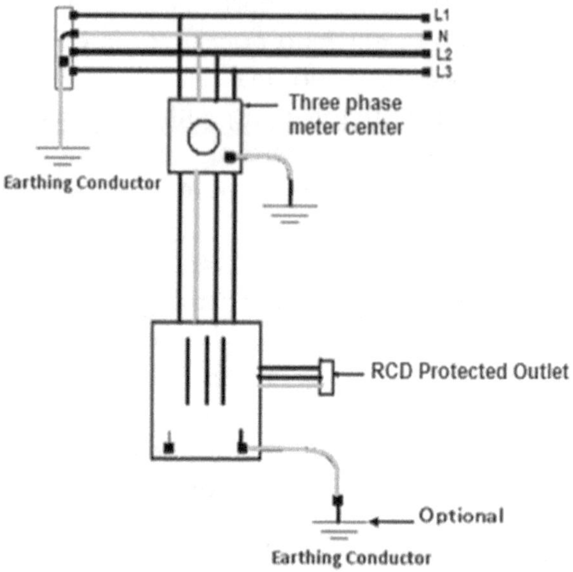

Figure 8.2. *TT system—150 kVA, 208 V, 3 Ø transformer supplied power with earth electrode at consumer's installation (no link between earth and neutral terminals)*

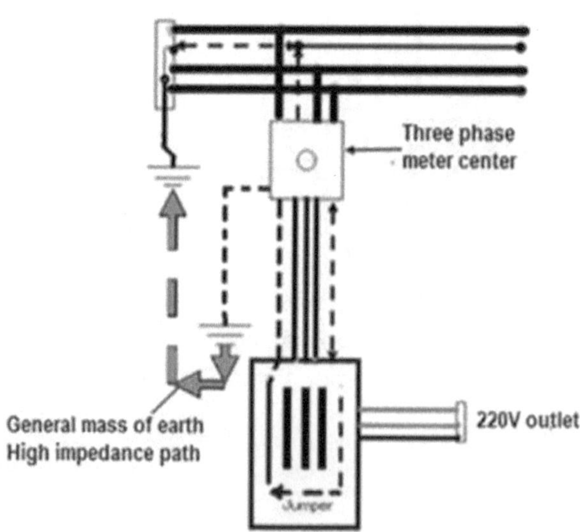

Figure 8.3: *TT and TNCS 150 kVA, 208 V, 3 Ø transformer combined earthing at a consumer's installation*

An essential bonding between neutral and earth in an electrical installation is shown in Figure 8.3. This is installed either at the meter center or at the distribution panel. It is preferred that this is installed at the meter center but not a combination of both. Bonding at the meter center is preferred as it eliminates the need for bonding at the distribution panels. Bonding at both points is not advised as this creates a parallel path or an arch ring which will not allow the dissipation of the fault current to the general mass of the earth. The neutral conductor has now become a low-impedance return path for short-circuit currents. It is critical that fault current returns to the neutral point of the transformer via the neutral conductor. In this earthing arrangement, the circuit breaker will operate quickly if a fault occurs.

8.2 Quick Trip Link (QTL)

The power available at the consumer distribution panel is proportional or equal to the power available at the transformer terminals. The available power is dependent on the size of the cable supplying the installation and the distance between the distribution panel and the supply transformer. It is for this reason that the link, *QTL,* between the neutral terminal and the earthing terminal is vital. This link creates a low-impedance path to the transformer's neutral, thus allowing for quick action when circuit breakers experience short-circuit currents.

The earthing configurations in the United States of America (USA), Canada, and the Caribbean use the combination of the TT and the TNCS systems as depicted in Figure 8.4. In the UK, Europe, and New Zealand, only the TNCS system is used. In the TT and the TNCS systems, the earth loop impedance (Zs) is essential, and emphasis must be placed on the lowest possible loop impedance which will cause circuit breakers to operate efficiently.

Zs is calculated by using the following formula:

$$Zs = Ze\,[(R1 + R2) \times Ca \times Ci]$$

Where *Zs* is the branch circuit impedance or the internal impedance

Ze is external impedance (transformer impedance and cable impedance to the main distribution board)

*R*1 is the resistance of the branch circuit phase conductor

*R*2 is the resistance of the branch circuit protective conductor (CPC) or equipment earthing conductor

Ca is the correction factor for ambient temperature

Ci is the correction factor for conductors embedded in thermal insulation

Note that the general earth fault loop impedance comprises the following:

1. The *branch circuit* (receptacle/socket outlet) *protective conductor* (earthing)
2. The consumer's earthing terminal and earthing conductor
3. The earth's return path via the earth electrode
4. The consumer's service conductor (neutral) from the transformer
   (Take note of the dotted lines and arrows in Figures 8.4 (A) and (B))

5 kVA, 208 V, 3 Ø

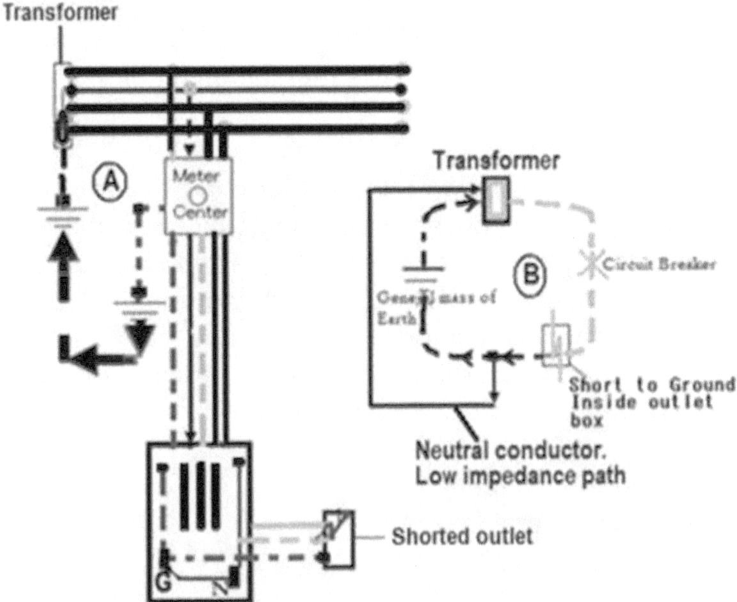

Figure 8.4: *The general earth fault loop*
(A) Installation of loop wiring (B) Loop illustration

8.3 Intra Circuit Earthing

In circuits where the neutral conductor is a part of the circuit, the neutral conductor provides the flexibility of creating the TNCS system at any point of the installation with maximum protection. This type of earthing arrangement at receptacles as shown in Figures 8.5 and 8.6 concentrates the fault current at the point where the fault is created.

The TNCS earthed neutral configuration provides essential protection against short circuits or transients at any point of the installation. This means, the neutral conductor in any installation will operate as a CPC without a physical earth conductor bonded in the circuit. This configuration will also provide maximum protection against fault current not large enough to cause a circuit breaker to operate. To provide maximum protection, all circuitry or equipment operating within this network must be protected by GFCI or RCD and tagged at the distribution panel, intra earth circuit. These devices when incorporated with the *intra earthing configuration* will protect users of appliance or equipment against direct and indirect contact with live metal framing of the appliance or equipment being used.

After carefully analyzing the earth fault loop impedance of Figures 8.4 (A) and (B), one will observe that electrical installations having met all mandatory circuit requirements *(earth electrode and low impedance link between neutral and earthing terminal)* regardless of the size of the installation. *Intra earthing configuration* will provide full protection for life and property against the following conditions: (See Figures 8.5, 8.6, and 8.7.)

1. Phase-to-earth fault
2. Phase-to-neutral fault
3. Transients
4. Lightning
5. Phase-to-phase

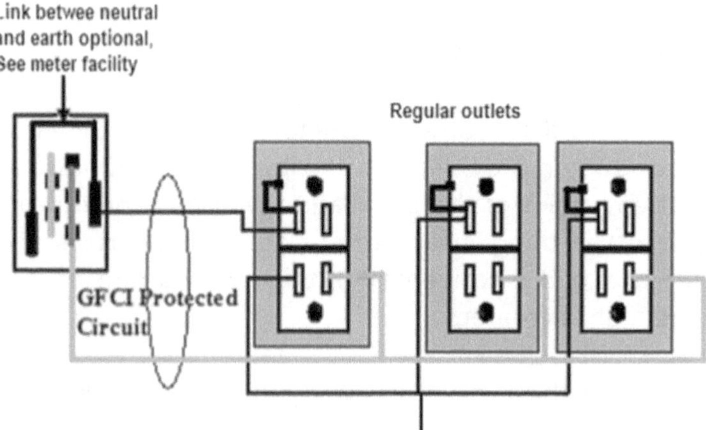

Figure 8.5: *Creating individual earthing protection at outlets for sensitive appliances or equipment using regular outlets*

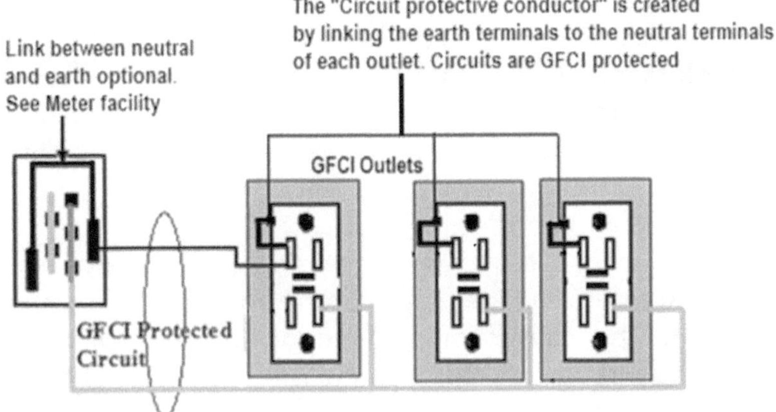

Figure 8.6: *Creating GFCI individual earthing protection at outlets for sensitive appliances or equipment*

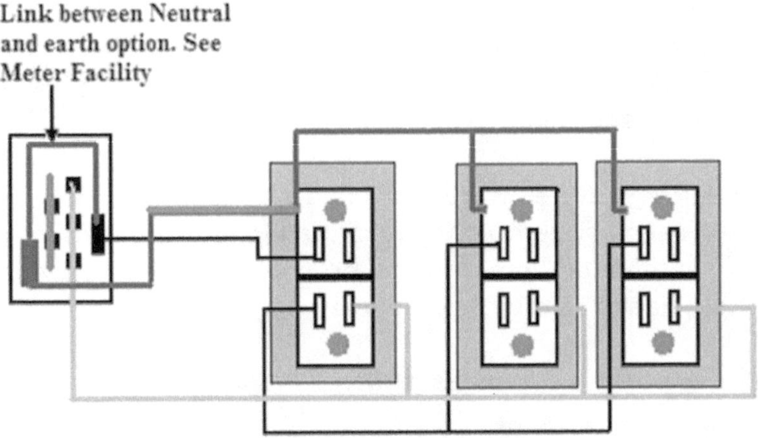

Figure 8.7: *Normal earthing configuration*

8.4 Clean Earthing System

Clean earthing system is a grounding configuration which is independent of the general earthing system of an installation. This system is free from external influences such as harmonics and transients.

Protection against electric shock in any installation is extremely crucial. Critically important is the need to pay keen attention to the earthing arrangements for television stations, transmission sites, processing plants, control systems, and information technology equipment which depend on proper earthing systems for their optimum performance.

Some malfunctions found in processing plant equipment are due to unwanted voltage and current appearing on or in the equipment. The causes of malfunction inclusive of mains or phase transients and earth transients may be due to any or a combination of the following:

1. Lightning
2. Load switching
3. Supply frequency
4. Earth differential voltages
5. Radio frequency fields
6. Magnetic fields

Clean earthing systems are not limited to processing plants and communication systems. They must be highly prioritized for areas where sensitive equipment is to be used, for example, studios, hospitals, health care facilities, and so on.

8.4.1 Creating a Clean Earthing System

Basic techniques used to achieve a clean earth and immunity to incoming electromagnetic disturbances are as follows:

1. Installing error correction equipment
2. Separating the equipment from the source electrically by installing an isolation transformer
3. Providing a low-impedance path to minimize the earth potential differences and providing shielding

Creating a clean earthing system is simply removing the ground connection between the MDP and the transformer and establishing an independent earth at the output of the transformer. This configuration prevents faults which are associated with the increase in neutral and ground current. It prevents fault currents from reaching the isolation transformer and from filtering into sensitive equipment. The increase in neutral and ground current could be due to lightning or other transients. Figure 8.8 shows an isolation transformer with a clean earthing system on the output.

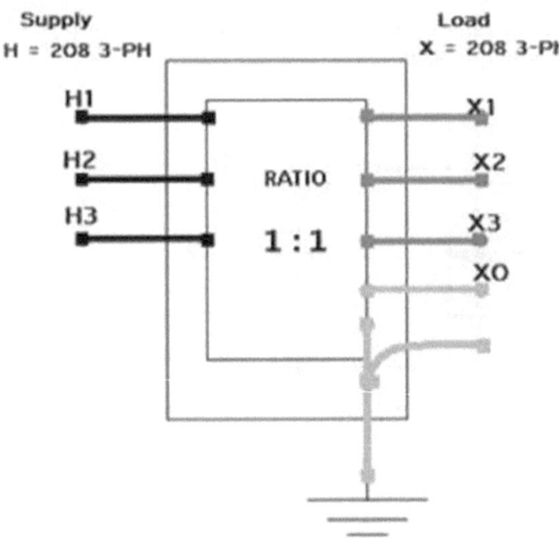

Figure 8.8: *Transformer which eliminates neutral or earth transients*

8.4.2 Neutral and Grounding Connections for Dry Transformers

There are two methods of configuring the main neutral and the main earth between the transformer and a distribution panel. Figure 8.9 shows the main earthing conductor connected from the transformer to the distribution panel. For this earthing arrangement, the neutral and the earthing terminal must not be linked; this may create a parallel path between the transformer and the distribution panel. Dedicated earth electrode or electrodes must be driven for the protection of the distribution panel. This will eliminate the traveling of earth potentials between the two points. The earthing conductor from this electrode must be connected to the main earthing terminal only as shown in Figure 8.9.

The second method of earth configuration between the transformer and distribution panel is shown in Figure 8.10. In this method, there is no earth conductor between the transformer and the neutral. Importantly, there must be an earth electrode tightly connected at the transformer neutral terminals and the earth terminal of the distribution panel. In addition, the neutral terminal and the earth terminal must be bonded together in the distribution panel.

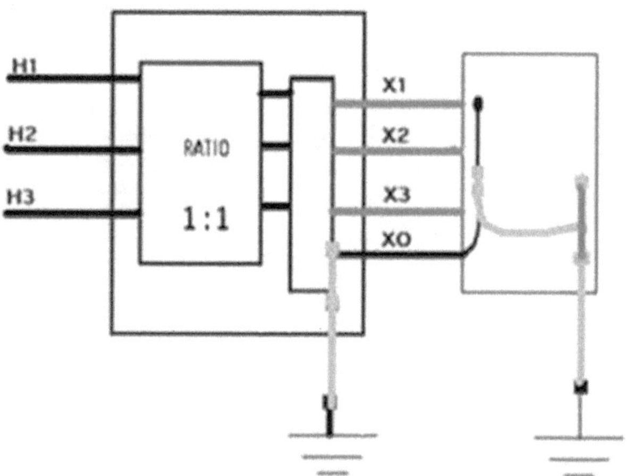

Figure 8.9: *Delta—Star: 208 V/208 V 3-phase isolation transformer*

Transients caused by load switching and the change of supply frequency are protected by transient suppression facilities which are fitted to the output of the transformer as shown in Figure 8.10. These components include, but are not limited to, CTs, contactors, potential transformers (PTs), and electrical devices.

Isolating transformers are ordered according to the sensitivity of the load they are intended to supply. Therefore, the type of load to be supplied must be obtained before ordering an isolating transformer.

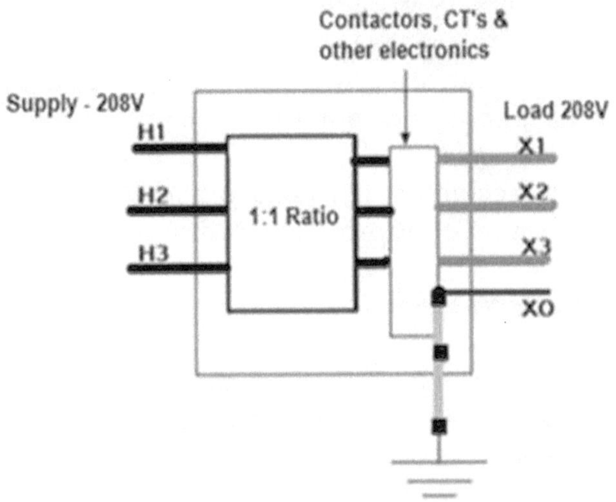

Figure 8.10: *Isolation transformer fitted with transient suppression facilities*

8.4.3 Transformer Neutral Points

Secondary neutral point (Xo) on a transformer must be bonded to the earthing terminal located in the distribution panel. As discussed in earlier chapters, bonding the earth to the neutral creates a low-impedance path via the neutral, which will allow the circuit breaker to operate in the event of a short circuit within 0.4 seconds.

The frame of the transformer must be bonded to Xo (secondary neutral point) of the transformer only. This bonding and grounding will provide the necessary earthing protection which is needed for all dry-type transformers.

8.4.4 Risk Assessment

Electricity and electrical-powered equipment have the potential of causing serious injuries or fatalities if associated risks are not fully mitigated. Persons using or working with electricity may not be the only ones at risk this risk is extended to all persons who are interfacing with the facility or equipment. Poor electrical installations and faulty electrical appliances can lead to serious injury, loss of life, fire and damage to equipment. Most electrical accidents can be avoided by careful planning and implementation of appropriate precautions.

8.4.4.1 Mitigating Electrical Risks

Actions to mitigate electrical risks are dependent on the situation that exists. The following list, though not exhaustive, will provide insights into some possibilities:

1. Choose equipment that is suitable for the intended working environment.
2. Use appropriate grounding configuration.
3. Avoid parallel path between neutral and ground conductors.
4. Use air, hydraulic, or hand-powered tools where appropriate.
5. Make sure that equipment is safe to operate and is fully maintained.
6. Provide an accessible and clearly identified switch near each fixed machine to turn off the power in an emergency.
7. For portable equipment, use socket outlets close by so that the equipment can be easily disconnected in an emergency.
8. The end of flexible cables should have the outer sheath of the cable firmly clamped to stop the wires (particularly the earth), pulling out their terminals.
9. Replace damaged sections of cable and ensure joints are sound.

10. Use proper connectors or cable couplers to join cables. Do not use strip connector blocks covered in insulating tape.
11. Protect light bulbs and other equipment which could be easily damaged during use.
12. In potentially flammable or explosive atmospheres, only special electrical equipment designed for these areas should be used. Seek a specialist's advice if necessary.

A risk assessment of a dry transformer is shown in Table 8.1. This displays the risks which are associated with connecting dry transformers and how to mitigate these risks. There is one question arising from the risk assessment: Without a ground or earth conductor connected to the transformer from the source as shown in Figure 8.11, would the circuit breaker supplying the unit operate if one of the transformer windings is burnt?

When a winding of a transformer is burnt, the impedance of that winding falls to an extremely low value. A transformer having a low impedance signifies that a short circuit is created, and the magnetic field ceases to produce. In fact, the burnt winding will operate as a piece of low resistance conductor. Hence, high current and LV will be experienced and cause the main circuit breaker to operate and isolate the transformer.

Colors used to indicate the various risk levels in Table 8.1 are as follows: :

- Very high chance of an incident (Red)
- Average chance of an incident (orange)
- Small chance of an incident (green)
- No chance of incident (light gray)

Table 8.1: Transformer risk assessment

Risk	Level	Mitigation	Level
Cable coming in contact with frame due to nicking of insulation	Red	Insulating all metal area cable passes through	Green
Phases coming in contact with each other	Red	Ensure power supplying the transformer is isolated or cut off before carrying out work	Green
Cable becoming loose due to heat and touching metals or other phases	Orange	Ensure cables are not under tension while being terminated Cables must be suspended freely inside terminals before tightening	Green
Burning of transformer windings	Red	Do not overload transformer above specification if transformer is used outdoor or in water area. Ensure transformer is Nema 4 or IP65 rated	Green
Circuit breakers in panel to which the transformer supplies do not operate under fault conditions	Red	Ensure that a link or jumper is installed between the neutral and the ground terminal at one point of the installation	Green

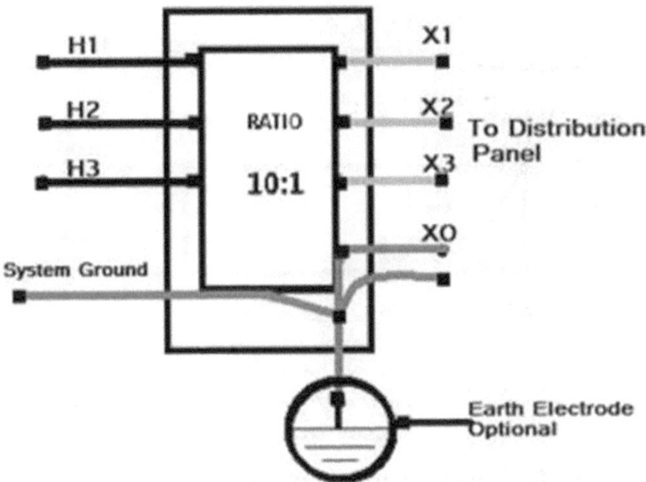

Figure 8.11: *Normal earthing of transformers (delta—star: 75 kVA, 208 V / 120 V 3-phase)*

8.5 Bonding a Clean Earthing to the Main System Ground

A clean earthing system must be bonded to the general earthing system to create an equipotential between all equipment and machineries in the facility. This bonding can be configured through one of the following two methods:

1. Bonding to an earth mat, plate, or grid, buried 3 feet in compact soil or the general mass of earth as shown in Figure 8.12 (commonly used in large industrial facilities).
2. Using a capacitor in the ground connection to provide a reciprocating fault current to solid grounding or to provide low or high frequency isolation (commonly used in large industrial facilities or for special facility equipment grounding).

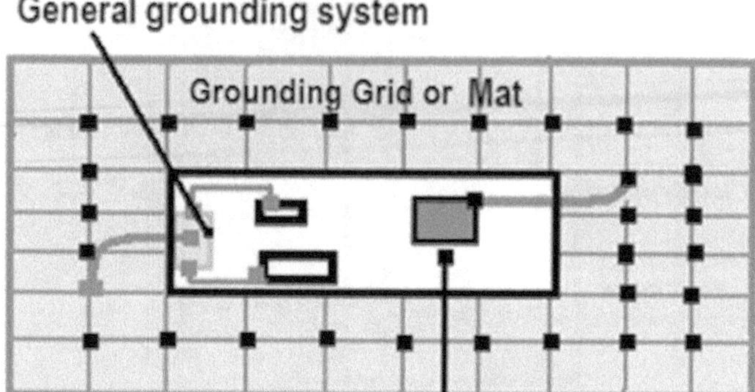

Figure 8.12: *Bonding to an earth mat, plate, or grid to reduce the effect of grounding frequencies*

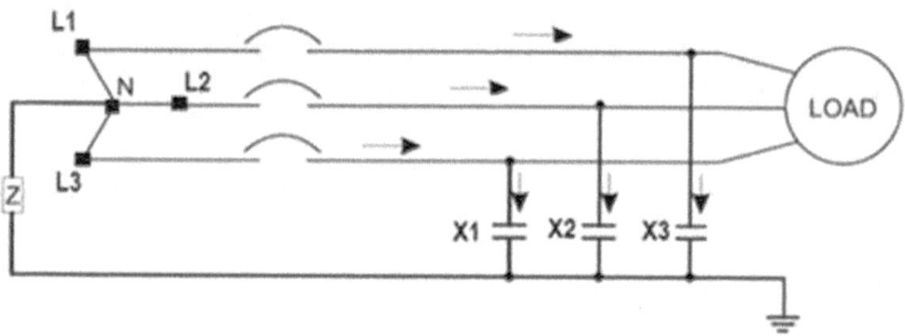

(Adapted from Littlefuse White Paper, 2010)

Figure 8.13: *Capacitive grounding*

The limit to practical low-level ground-fault protection in industrial electrical systems is a function of physical parameters. Current sensing is the best method to detect and locate ground faults; however, system capacitance, unbalanced loads, current-sensor limitations, and harmonics affect current measurement and limit the lower level of practical ground-fault detection.

Capacitive grounding shown in Figure 8.13 could be implemented as a solution to mitigate problems which may arise as a result of ground faults.

8.6 Conclusion

The type of grounding configuration proposed for an installation must take into consideration the conditions where the grounding is required. The multiple earthing system is one of the most commonly used configurations throughout the world. This type of configuration has been proven to be very effective. Understanding the behavior of fault current allows the technicians to manipulate the use of earthing configuration to suit their needs.

Elimination of fault current from sensitive equipment is essential. Installations such as television stations, transmission sites, processing plants, control systems, and information technology equipment depend tremendously on proper grounding for their optimum performance.

Creating a clean earthing system is simply separating the utility neutral and ground system from the equipment to be supplied. This is done by creating an independent neutral and grounding system on the secondary side of a 1:1 ratio transformer. Thus, a clean neutral and earth system free of external contamination and faults will be available to supply sensitive equipment where required. These types of grounding equipment are available in ranges from small kW ratings to large kVA ratings.

8.7 Test Your Knowledge

1. Define Zs.
2. Calculate the Zs of a circuit where *Ze* = 0.3 Ω, *R2* = 0.6 Ω, and *R1* = 0.9 Ω. Ignore the correction factors in this instance.
3. Explain what is meant by TT and TNCS.
4. Sketch the earthing arrangement for a TT and TNCS configuration.
5. Explain the behavior of fault current in a TT and TNCS earthing arrangement.
6. Sketch a simple diagram showing the general Zs.
7. Explain the importance of bonding the main earth terminals and the main neutral terminals at the meter center.
8. List four faults that an installation should be protected against.
9. Draw an earthing arrangement which will protect an individual circuit from faults which may occur on other circuits.
10. Explain the effect of a high or low circuit impedance on a circuit experiencing a short circuit.
11. Name two types of transient which will put severe pressure on an earth electrode.
12. Sketch two types of intra earth configuration.
13. Explain the importance of a GFCI outlet or circuit breakers used in an intra earthing circuit.
14. Sketch two simple diagrams showing the multi-earthing of a domestic service line with three poles between the meter facility and the distribution panel and a transformer pole supplying five other distribution poles.
15. What is the purpose of a multi-earth system?
16. What is meant by "a clean earthing system"?
17. Describe the principles of operation of an isolation transformer.
18. Name two types of transients.
19. Name four causes of transients.

20. Name two types of equipment used to mitigate transients.
21. Why is it important not to connect the earth or neutral from a supply panel to the secondary side of an isolation transformer?
22. What is the winding ratio used to construct isolation transformer?
23. Explain why is this type of ratio winding critical?
24. Draw and label two types of grounding connections which can be made on a dry transformer.
25. What is the purpose of current transformers C.T's and potential transformers P.T's in an isolation transformer?
26. Name five institutions where the use of isolation transformers should be considered.
27. Explain what is meant by risk assessment.
28. Do a risk assessment of the grounding connection to be made on a dry transformer to supply sensitive equipment.
29. Show two method of linking the neutral of a transformer its protective grounding.

Bibliography

i. HSE, Health and Safety Executive. April 2012. Electrical Safety and You. http://www.hse.gov.uk/pubns/indg231.pdf

ii. The Institute of Electrical Engineers—IEE. 2008. Requirements for Electricians. Wiring Regulations. Sixteenth ed. Pages 46-47.

iii. The Institute of Electrical Engineers—IEE. 2002. *On-site Guide*, Sixteenth ed. Page 133

iv. The Institute of Electrical Engineers—IEE. 2002. *On-site Guide*. Sixteenth ed. Appendix 9 Page 157

v. Littelfuse: White Paper, Ground Fault. 2010. Lowering the Limits for Ground Fault Detection. Retrieved on November 3, 2013, from http://www.littelfuse.com/products/relays-controls-and-systems/protection-relays/~/media/Files/Littelfuse/Technical%20Resources/Documents/White%20Papers/Littelfuse-Protection-Relay-Lowering-the-limits-for-ground-fault-detection.pdf

vi. Schneider Electric. Authorized Person (HV Electrical). June 2003. *Training Manual*

CHAPTER 9

Fault Current Management

9.0 Introduction

Fault current management is the control of fault or unintentional voltage and current, which may be a threat to life and/or property. Fault current management significantly relies on three elements addressed earlier in this book: earth electrode, soil management, and equipment bonding. As discussed earlier, the use of an earth electrode is to create a low-impedance path between the installation and the general mass of earth. This plays an important role in the control or management of fault current.

9.1 Soil Resistivity and Fault Current

In most cases, the *impedance* of the general mass of earth is unacceptably high; consequently, the earth or soil return path via the general mass of earth to the transformer will not be sufficient to ensure that maximum short-circuit current will flow to allow the protective device to operate. Hence, the meaning of soil resistivity becomes applicable.

> *Soil resistivity is the measure of the resistance between the opposite faces of a cube of homogeneous soil measured in Ω/m, or Ω/cm.*

Similar to soil's resistance, appliances and equipment which have metal frames and are within reach of each other and have differences in resistance and potentials could be fatal if not properly managed during a fault to earth.

For example in one particular case, an employee, Mr. Gerard Copper worked at a concrete block factory where he was required to work between two block, making machines at the same time. Over a number of months, these machines were in full operation without having had a thorough electrical inspection, testing, or maintenance.

On one afternoon, Mr. Cooper was on his normal routine when suddenly he was held by a fault current that had developed between the machines, shown in Figure 9.1, which had two different potentials. Mr. Cooper's excruciating cries caught the attention of his co-worker, Mr. Nad, who rushed to his assistance and quickly disconnected the main switch of the machine supply. Mr. Copper barely missed death and was able to recuperate in the hospital.

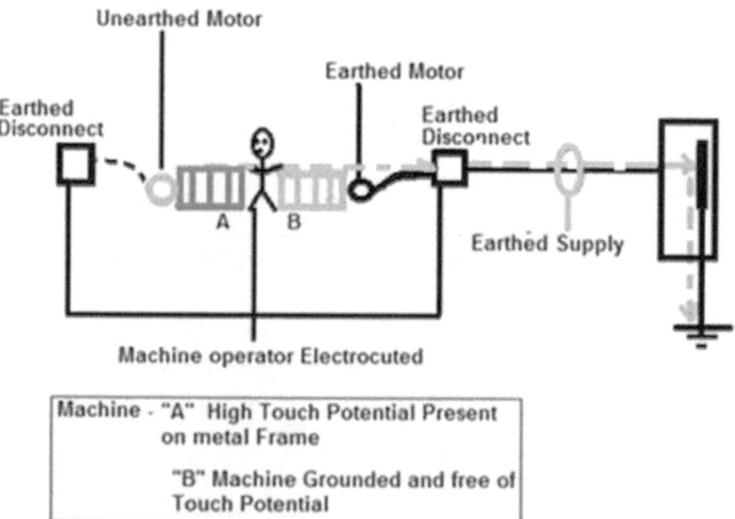

Figure 9.1: *One of two machines within reach of each other, carrying dangerous fault current*

To eliminate the occurrence of dangerous fault current as shown in Figure 9.1, it becomes critical to manage the flow of fault current through the process of bonding. Bonding of an electrical installation is simply a connection between all points of the installation (lighting and power circuits, fixed appliances or equipment circuit, etc.) and one common point, namely the main neutral or main earthing terminal of the installation.

Creating bonding between all metal parts of an installation causes the touch potential of all points of the installation to be the same. Bonding as shown in Figures 9.2, 9.3 and 9.4 provides protection against electrical shocks or electrocution. Proper bonding techniques in an installation will also ensure that the magnitude of current and voltage to earth during a fault is kept within predictable limits.

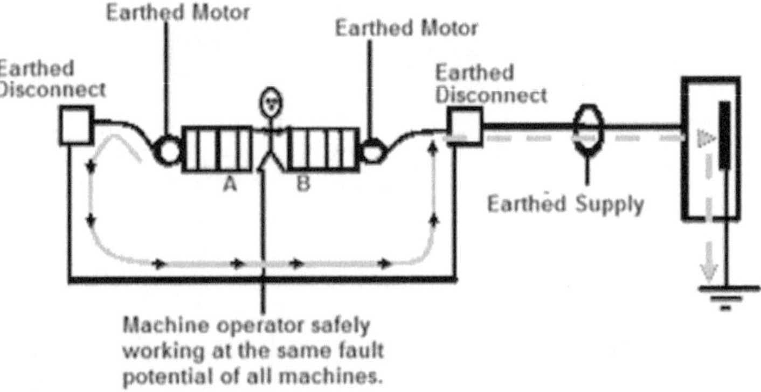

Figure 9.2: *Bonding all metal parts to reduce the risk of electrocution*

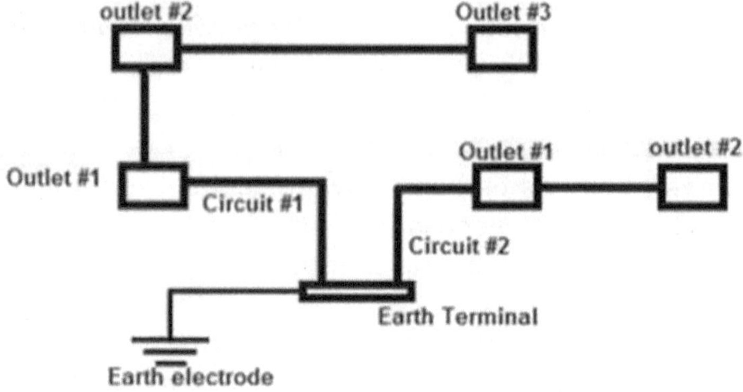

Figure 9.3: *Bonding two circuits to a common point*

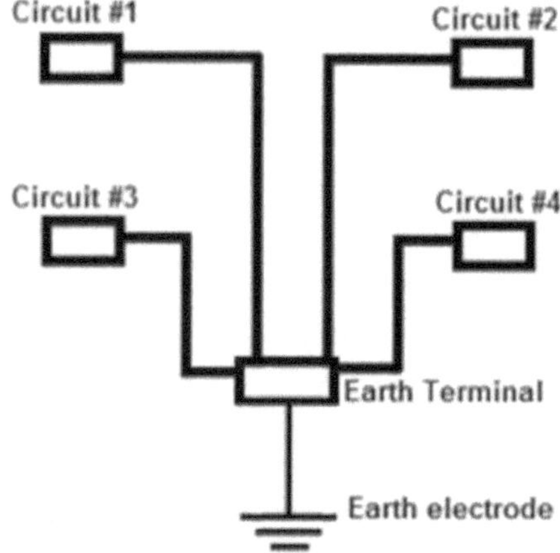

Figure 9.4: *Bonding individual circuits*

9.2 Determining the Size of an Earth Electrode

The size of an earth electrode to support an installation can be determined if certain parameters are known. These include the maximum current to be dissipated in the earth, the resistivity of the soil, and the time required for the current to be dissipated. For example, if the maximum current to be dissipated to the earth is 10,000A, the soil resistivity is 8 Ω/m, and the time taken to dissipate the current is 0.8 seconds, determine the size of the earth electrode required for soil with 30% moisture.The following steps can be followed to determine the size of the earth electrode:

Step 1: Determine the current density

$$\textbf{Current Density} = \sqrt{\frac{\mathbf{I}}{\mathrm{P\,x\,t}}}\ \mathrm{A/m^2}$$

Where *I* is the current flow to the general mass of earth if a short circuit occurs on the installation. *P* is the soil resistivity. *t* is the time which the current takes to dissipate in the general mass of earth.

$$I = \sqrt{\frac{10{,}000}{8 \text{ x } 0.8}}$$

$$I = 40 \text{ A/m}^2$$

Therefore, the current density (*I*) for clay with a resistivity of 8 Ω/m and 30% moisture is 40 A/m^2.

Step 2: Determine the electrode surface area for specified fault current

If a fault current of 10 A exists on an installation, what size electrode would be needed to dissipate the current and reduce the touch voltage and current on that installation?

Determine the surface area of the electrode required to dissipate the fault current.

$$\textit{Electrode surface area} = \frac{\textbf{Fault Current}}{\textbf{Current Density}}$$

$$= 10 \text{ A} / 40 \text{ A/m}^2$$

$$= 0.25 \text{ m}^2$$

The electrode surface area needed to be buried to dissipate 10 A is 0.25 m^2.

We know the surface area required can now be determined by the length of the electrode which will be equivalent to the surface area of 0.25 m^2.

Step 3: Given the diameter and length of a standard earth rod, determine the number of rods that are required.

Let us look at a standard length of electrode (5/8" diameter and 8 ft. in length).

We must first calculate the area of the rod by using the formula

Area of rod = $2\pi r\,(r + h)$
Radius of rod =0.3125 in, converted to meter = 0.008 m
8 ft convert to meter = 2.4384 m
Surface area of rod = 2 x 3.142 x 0.008 x (0.008 + 2.4384)
= 0.05 x 2.45
= 0.123 m^2

The number of electrodes required is:

$$= \frac{\textbf{Surface Area required to dissipate the fault current}}{\textbf{Surface area of electrode chosen}}$$

Number of 8 ft. electrodes required = 0.25 m^2 / 0.123 m^2

= 2

Two (2) 8 ft rod will be required to dissipate a fault current of 10 A, providing that the installation uses the multiple earthing system (TNCS).

Example 9.2 below analyzes a rocky soil condition and the length and number of the earth electrode required, having a resistivity of 600 Ω/m in comparison of the previous example with a resistivity of 8 Ω/m.

Example 9.2

Soil resistivity = 600 Ω/m

$$I = \sqrt{\frac{10,000}{600 \text{ x } 0.8}}$$

$$= \sqrt{\frac{10,000}{480}}$$

$I = 4.6 \text{ A/m}^2$

Therefore, the current density (I) for clay with a resistivity of 600 Ω/m and 30% moisture is 4.6 A/m^2.

If a fault current of 10 A exists on an installation, what size electrode would be needed to dissipate the fault current of 10 A and reduce the touch voltage and current on that installation?

First, find the surface area of the electrode required to dissipate the fault current.

$$Electrode\ surface\ area = \frac{\textbf{Fault Current}}{\textbf{Current Density}}$$

$= 10 \text{ A} / 4.6 \text{ A/m}^2$

$= 2.17 \text{ m}^2$

The electrode surface area needed to be buried to dissipate 10 A is 2.17 m^2

Since we now know the surface area required to dissipate the fault current, we can now determine the number of 5/8″ diameter 8′ rods equivalent to the surface area of 2.17 m^2.

Radius of 5/8″ rod = 0.3125″ When converted to meter = 0.008 m

8 ft converted to meter = 2.432 m

Area of rod = $2\pi r\,(r + h)$

$$= 2 \times 3.142 \times 0.008 \times (0.008 + 2.432)$$
$$= 0.05 \times 2.45$$
$$= 0.123 \text{ m}^2$$

The number of electrodes will be

$$= \frac{\textbf{Surface Area required to dissipate the fault current}}{\textbf{Surface area of electrode chosen}}$$

Number of 8 ft.
electrodes required = 2.17 m^2 / 0.123 m^2
= 17

In this location, with a resistivity of 600 Ω/m the number of 8 ft rods required is 17.

Examples 9.1 and 9.2 show that more earth electrodes of similar length will be required in an area with lower current density for a similar fault current. Longer rods could be used to reduce the number of rods required. It is for this reason that the bonding of the main earth and neutral bar located at the MDP become critically vital.

Notwithstanding the number of electrodes needed to hold a steady fault current of 10 A to ground, one eight feet (2.432 m) rod is adequate to provide protection to life and property for small and medium installation (domestic and /or commercial) because of the bonding between the neutral and the earth terminals as stipulated by regulations.

According to the NEC, an installation is safely connected to the general mass of earth with an earth resistance of 25 Ω. This resistance is dependent on the soil type of the immediate area. However, due to the variants in the composition of soils, huge differences in soil resistivity will be found in many areas and may result in the use of several earth electrodes to achieve the required earth resistance reading.

Calculating the length for earth electrodes will not necessarily coincide with standard lengths, consequently the nearest standard length above that which is calculated is employed. Standard lengths for electrodes include 4ft, 6ft, 8ft, and 10ft. It has been established that these standard rods are adequate for domestic installations and some commercial buildings in areas where the soil resistivity is exceptionally low. A minimum of two electrodes should be used in areas where the resistivity is extremely high. See Tables 9.1 and 9.2.

Table 9.1: Soil resistivity values and expected resistance reading of bare copper and rods buried and driven in the related soils

Soil/Ground Type	Typical Resistivity (Ωm)	For homogenous soil likely resistance of a 2.4m rod (Ω)	For homogenous soil likely resistance of 50m of a 70mm² earthwire(Ω)
Mercia Mudstone	20	8	0.8
Coal Measures	20	8	0.8
Loam	25	10	0.9
Alluvium	35	14	1.3
Boulder Clay	50	20	1.9
Keuper Marl & Waterstones	50	20	1.9
Head	70	28	2.6
Sand/Gravel	300	120	11
Limestone	300	120	11
Pebble Beds	300	120	11
Permian Limestone & Marl	400	160	15
Gritstone	1000	400	38

(Adapted from Earth Manual Section E3, December 2006)

Table 9.2: Soil resistivity values and expected resistance reading of earth electrode planted in a mixture soils

Soil Description	Group Symbol	Average Resistivity (Ω-cm)	Resistance of 5/8 in x 10ft Rod (Ω)
Well grade gravel-sand, gravel-sand mixture, little or no fines	GW	60,000 - 10,000	180 - 300
Poorly graded gravels, gravel-sand mixtures, little or no fines	GP	100,000-250,000	300 - 750
Clayey gravel, poorly graded gravel, sand-clay mixtures	GC	20,000-40,000	60 - 120
Silty sand, poorly graded sand-silts mixtures	SM	10,000-50,000	30 - 150
Clayey sands, poorly graded sand-clay mixtures	SC	5000-20,000	15 - 60
Silty or clayey fine sands with slight plasticity	ML	3000-8000	9 - 24
Fine sandy or silty soils, elastic silts	MH	8000-30,000	24 - 90
Gravelly clay, sandy clays, silty clays, leam clay	CL	* 2500-6000	*17 - 18
Inorganic clay of high plasticity	CH	* 1000-5500	*16 - 16
* These soils classification resistivity results are highly influenced by the presents of moisture			

(Adapted from Earth Manual Section E3, December 2006)

9.3 Soil Factor

The condition of the soil in which a fault is to be dissipated significantly affects the rate at which the dissipation of current and voltage takes place during a fault or a short circuit. In addition to a normal fault, transients such as lightning use the same grounding path to dissipate high current or voltage. It is for this reason that carefully designed and managed current path is crucial for all installations. A perfect connectivity to the general mass of earth is influenced by the following conditions:

1. **The soil impedance or resistivity**
   Influences the speed at which the current dissipates to earth.
2. **The soil quality**
   Influences the impedance of the general mass of earth and the area of dissipation

3. **The moisture content**
   Influences the bonding of the soil and qualifies the general soil impedance
4. **The soil temperature**
   Influences the resistivity of the soil and dictates the soil impedance

These conditions in coordination control the speed and the magnitude of any dissipating fault current to the general mass of earth.

9.4 Soil Types

Soil characteristics and conditions of each area vary due to the underlain and the overlain soils and the moisture content in each area. It is for this reason the value of resistivity will change from place to place. Usually, in low-lying areas, the overlain soil type is clay, and in hilly areas, the overlain soil type is rocky, stony, or of a limestone composition with a small amount of overlain soil.

In addition, the type of earthing arrangements may differ to facilitate the soil quality for each area, namely the length and number of earth rods or electrodes. In hilly areas, two or three earth electrodes may be required to achieve an acceptable earth resistance reading. This also depends on the depth of the topsoil, the moisture levels of the soil, the corrosiveness of the soil, and the mixture of the soils.

Areas in countries such as Jamaica are categorized by predominant soil type, and they are as follows: (See Figures 9.5 and 9.6.)

1. JM1—Brownish yellow clay
2. JM2—Reddish yellow stony clay
3. JM3—Intense red clay predominant in clay minerals and fraction of gibbsite
4. JM4—Red clay
5. Blue Mountain—Stony and shallow clay

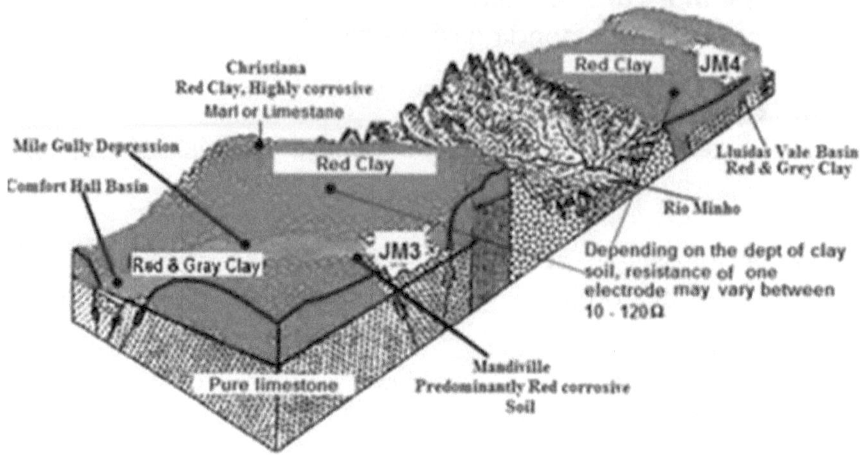

(Adapted from Soil Briefing Jamaica, August 1, 1995)

Figure 9.5: *Typical land geological structure and likely earth electrode resistance readings in Jamaica—areas JM3 and 4*

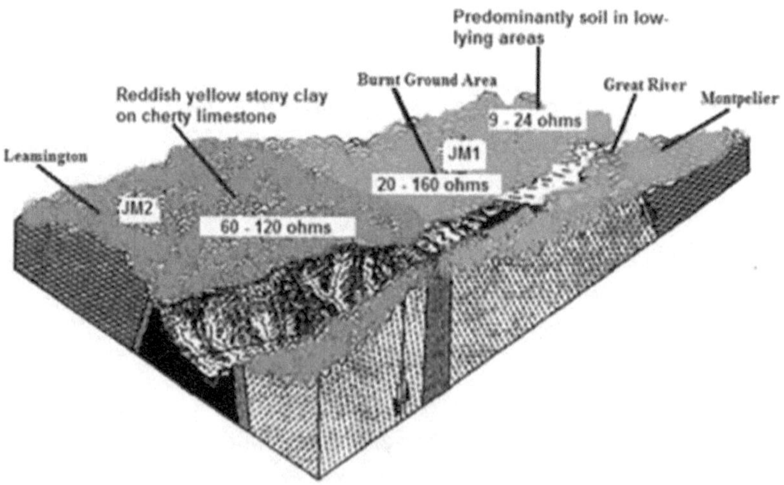

(Adapted from Soil Briefing Jamaica August 1, 1995)

Figure 9.6: Typical land geological structure and likely earth electrode resistance readings in Jamaica—areas JM3 and 4 (coastal and low-lying areas are excluded).

9.5 Corrosiveness of Soils

Some soils have an excessive amount of chemicals and salts which significantly affect some metals buried underground. These soils are termed as aggressive soils. This means soils which are aggressive and are also highly corrosive. This aggressive condition does significant damage to metals which are unable to withstand corrosive action. Copper, silver, and stainless steel are almost invulnerable to corrosion, and they are compatible with almost all underground conditions.

Copper is the least expensive of the three metals and is the most widely used electrical conductor. The compatibility of this metal to all soils is due to the electrolytic refining process which eliminates the impurities such as phosphorus, arson, iron, titanium, and silicon which would cause copper to corrode and dissolve quickly in soils.

Corrosion of metals buried underground occurs when an above-normal level of sulphate or chloride in conjunction with high moisture content is present in the soil. Corrosion could also occur where soil resistivity is found to be extremely low. This could suggest that the low soil resistivity found could be a result of a high level of salt or chemicals present in the general area of the soil. Salt content in soils is a significant factor in relation to low resistivity. Soils with salt content above 1% by weight keep soil resistivity levels low and constant. Soils with significantly less salt content (below 1%) will be found to have high soil resistivity as shown in Table 9.3.

Generally, homogeneous soils such as clay, sand, gravel, and loam do not possess properties that are associated with corrosion. However, because of the unpredictable aggressive condition of soils, galvanized rods should not be used as earth electrodes.

Table 9.3: The effect of salt content on the resistivity of soil

The Effect of salt content on the resistivity of soil (Sandy loam, Moisture content, 15% by weight, Temperature, 17°C)	
Added salt (% by weight of Moisture)	Resistivity (Ohms-centimeters)
0	10,700
0.1	1,800
1.0	460
5	190
10	130
20	100

(Adapted from Why Measure Soil Resistivity? August 2002)

9.6 Moisture and Temperature

The resistivity of the general mass of earth is affected by the moisture levels, temperature, and chemical or salt content of the soil. Furthermore, resistivity readings of soils are more likely to change if they are dampened by rain, river, irrigation, or seawater. In addition, the sun's penetration of the general mass of earth will affect change in resistivity of the general mass of earth.

Resistivity increases where temperatures are found to be below 20° C and remains constant where temperatures are above 20° C as shown in Table 9.4. It is for this reason that soil resistivity is being classified as seasonal.

Soils with a high percentage of moisture offer less impedance to the flow of fault current in the general mass of earth, and the reverse is true if the soil has a low percentage of moisture. Table 9.5 represents the resistance of different types of soil at various moisture levels and temperatures.

Table 9.4: Variation in resistivity with temperatures for a mixture of sand and clay with a moisture content of about 15% by weight

Temp. C	*Typical resistivity Ωm*
20	72
10	99
0 (water)	138
0 (ice)	300
-5	790
-15	3300

(Adapted from Earthing Techniques, n.d.)

Table 9.5: Soil resistivity and moisture levels

Type of Soil or Water	*Typical Resistivity Ωm*	*Usual Limit Ωm*
Sea water	2	0.1 to 10
Clay	40	8 to 70
Ground well & spring water	50	10 to 150
Clay & sand mixtures	100	4 to 300
Shale, slates, sandstone etc	120	10 to 100
Peat, loam & mud	150	5 to 250
Lake & brook water	250	100 to 400
Sand	2000	200 to 3000
Moraine gravel	3000	40 to 10000
Ridge gravel	15000	3000 to 30000
Solid granite	25000	10000 to 50000
Ice	100000	10000 to 100000

(Adapted from Earthing Techniques, n.d.)

9.7 Let-through Current of Soils

The disconnecting time for a short circuit on a TT system combined with TNCS system averages 0.8 sec. The TT earthing arrangement of itself will not allow a quick disconnecting action of circuit breakers under short-circuit conditions even if the soil type for the installation were loam with a resistivity of 208 Ω/m as determined in Example 9.3.

If the soil were to be used as the only fault return path, the time which a fault current takes to pass through the earth electrode, to the general mass of earth and to the utility transformer, equals or greater than the time (t) which the circuit breaker takes to operate. Additionally, if the let-through fault current is sufficient, significant damage could be done to property by the fault current in this prolonged period and could lead to fatalities.

Figure 9.7 illustrates the dissipation of current in soils and emphasizes the importance of earth electrodes to be properly connected with the general mass of earth.

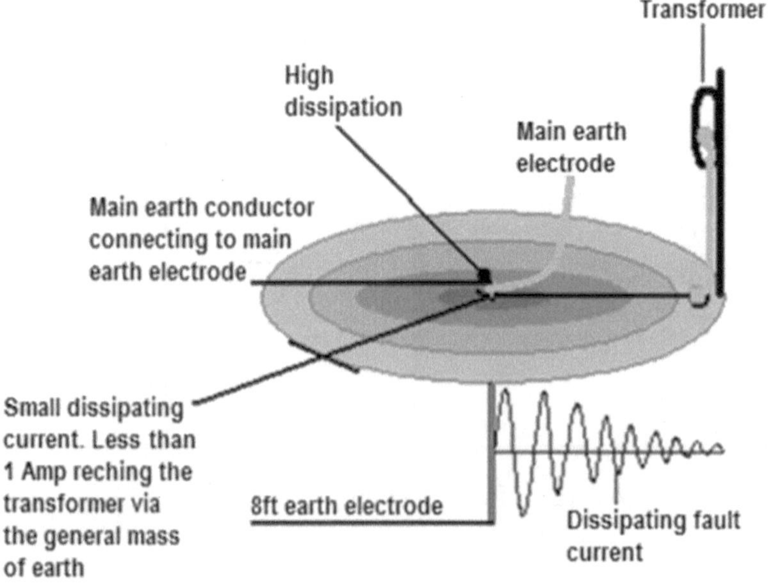

Figure 9.7: *Current dissipating in soils*

Quick dissipation of transients or fault current requires that the grounding or earthing of an electrical installation must have the correct length and number of earth electrodes or rods in order to achieve maximum connectivity to the general mass of earth. This connectivity will provide a low-impedance path, allow full dissipation of fault current to the general mass of earth, and improve connectivity between all metal parts of the installation and the general mass of earth.

In addition to the dissipation of current, where achievable, all earth electrodes should be placed at a minimum of 10 feet away from a structure or building as shown in Figure 9.8. This distance will prevent dangerous potential from high transient's currents or direct lightning charges, entering structures or buildings via the general mass of earth.

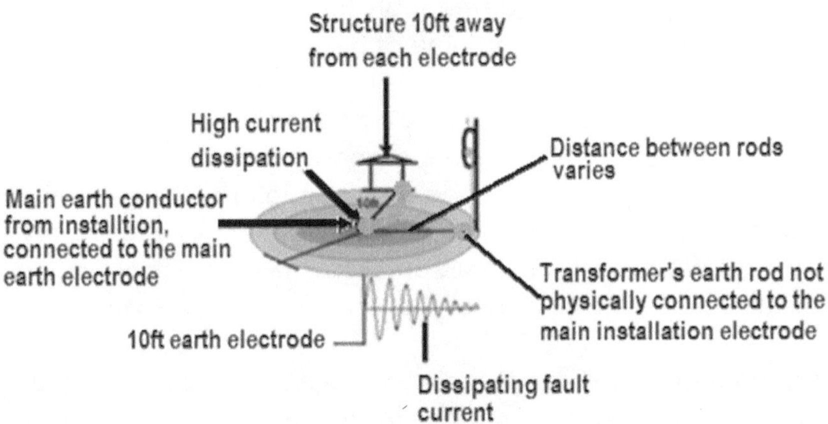

Figure 9.8: *Placement of earth rod (10 feet from building)*

Example 9.3

A fault current of 100 A with a voltage of 2 V was injected into the general mass of earth. What is the let-through current if the soil resistivity is 300 Ω/m and the installation is located 35 m from the distribution transformer?

The let-through current of the soil per meter

$$I_{let.\ thr} = \frac{V}{R}$$

$$= \frac{2}{300} = 0.0067 \text{ A/m}$$

From the point of injection of 100 A at the earth electrode to the distribution transformer located 30 m away from the electrode, the let-through current would be

Idist. Tran. = let-through current per meter x Distance between transformer and electrode
= 0.0067 × 30 = 0.2 A

This current would not be enough to cause a circuit breaker to operate within the required time of 0.4 second.

Although the fault current of 0.2 A reaching the transformer would not be high enough to initiate the operation of the circuit breaker, the fault current of 100 A will be seen as a load to the general mass of earth and an overload to the circuit breaker. As a result, instant or monetary tripping of circuit breakers may occur, depending on the value and the resistivity of the general mass of earth.

The monetary tripping of a circuit breaker due to a short circuit may result in electrical shocks from metal frames of electrical appliances. If these appliances are not properly bonded to the electrical system, it is likely to cause electrocution. Furthermore, if an appliance is in close proximity to another appliance with fault current on its frame and without the necessary bonding, a different potential will be present between both appliances, posing a high risk of electrocution as shown earlier in Figure 9.1.

Since the dissipation fault current to the general mass of earth may be extremely slow as a result of the type of soil as shown in Figure 9.9, the fault current may become a risk to life and property.

Grounding tape or a number of electrical rods may be used to significantly reduce or eliminate the risk of electrocution or electric shock in these circumstances.

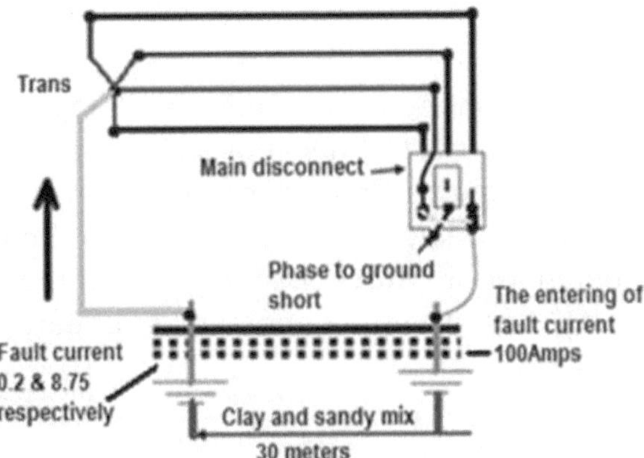

Figure 9.9: *100 A dissipating in a clay sandy mixture soils*

Example 9.4

A fault current of 100 A with a voltage of 2 V was injected into the general mass of earth. What is the let-through current if the soil resistivity is 8 Ω/m and the installation is located 35 m from the distribution transformer?

The let-through current of the soil per meter is:

$$I_{let.\ thr} = \frac{V}{R}$$

$$= \frac{2}{8} = 0.25\ A/m$$

From the point of injection of 100 A at the earth electrode to the distribution transformer located 30 m away from the electrode, the let-through current would be

*I*dist. Tran. = let-through current per meter x Distance between transformer and electrode

= 0.25 × 30 = 7.5 A

From the point of dissipation of 100 A to the distribution transformer, the let-through current via the general mass of earth would be 7.5 A. This current would not be enough to cause a circuit breaker to operate by a short circuit within the required time.

Electrocution is imminent if the main earthing conductor or earthing lead is disconnected and the soil is the only protection for the electrical system. Removing this connection removes the only safe path for fault current to the general mass of earth. Therefore, human or livestock coming in contact with these points will close the path to the general mass of earth as shown in Figures 9.10 and 9.11.

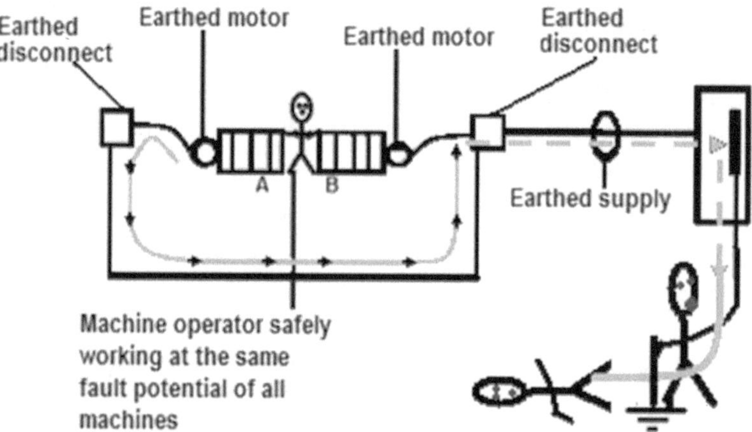

Figure 9.10: *Disconnecting an earth conductor without testing could be fatal*

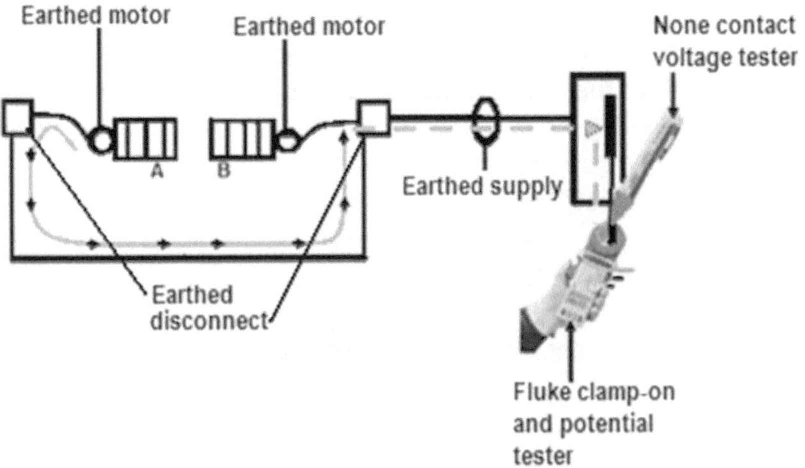

Figure 9.11: *Critical testing before disconnecting or troubleshooting grounding systems*

9.8 Grounding Poles in Lightning-prone Areas

Fault current management of distribution poles plays a key role in the amount of fault current diverted to the consumer's premises. Most fault current is distributed through neutral conductors.

All utility poles used for transmission and distribution lines, conductive or non-conductive, shall be properly grounded and bonded to all neutral conductors to improve the efficiency of the distribution system ground neutral network in all terrains, rocky or clay. Poles where possible should be grounded to facilitate the quick dissipation of all neutral/ground fault current to the general mass of earth.

Transformer poles may sometimes be several pole spans away from where a fault current or a lightning transient is initiated. In these instances, service laterals or service equipment within the fault path may experience high fault current, causing collateral damage to equipment or any electrical component on the transformer network.

Where the fault current is too large and does not dissipate quickly enough to the general mass of earth, the transformer fuse should operate to clear the fault. This fuse operation may not take place if the transformer upstream is located at a significant distance from where the downstream fault occurs. Therefore, distribution lines should be designed to ensure that the last pole and mid poles of service laterals are able to return a fault current large enough to allow the transformer fuse to operate.

9.9 Impedance of Service Lines

Low short-circuit current found on distribution systems could be due to high line impedance due to distance, size, or tightly bunched conductors in the same conduits. Tightly bunched conductors in the same conduit will cause an increase in temperature which directly affects impedance. Therefore, the management of conductor sizes and the bunching of cables into conduits or otherwise must be given particular attention in all installations. Giving significant attention to these areas increases the operating efficiency of all protective devices.

If a voltage drop below the standard requirement of 10% is present on an installation, it suggests that high line impedance is present in the affected section of that installation. This means, protective devices could take a longer time to operate under fault conditions.

Fault current on a domestic or commercial installation as a result of short circuits, namely *live-to-neutral* and *phase-to-phase,* should operate within 0.8 sec of the time the fault is created to protect life and property. With current sensitive technological devices, 5 seconds is enough time to cause catastrophic damage.

9.10 Dissipation of Fault Current to Earth

Fault current and voltage as a result of major short circuit or lightning is broken up into pieces as they travel along the distribution lines by the multi-earthing system of the distribution network as shown in Figure 9.12. As a result, HV and current will be present on the earthed neutral of the distribution network. It is for this reason that the earth electrode at transformer poles or dry transformers should have the lowest achievable resistance to earth; this will provide a quick dissipation path for high fault current or transients to the general mass of earth.

If eight poles as shown in Figure 9.13 are on a distribution network with earth electrode or earth resistance reading 5 Ω for each electrode. Using Ohms law, this will result in the network having a combined reading of 0.625 Ω. It cannot be overemphasized that more than one earth electrode at any one point of an installation as shown in Figure 9.14 will increase the connectivity of that installation with the general mass of earth. This will reduce the resistance to earth and create a path for quick dissipation of fault current to earth.

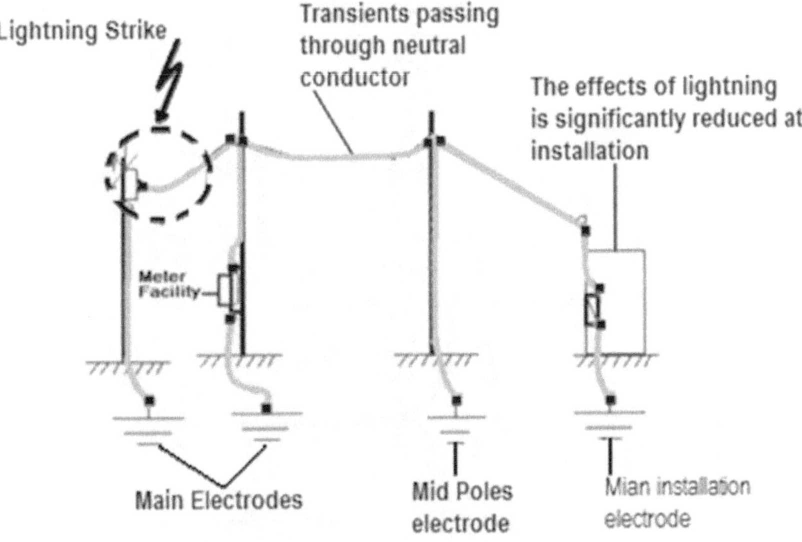

Figure 9.12: *Multiple earth electrodes to improve the dissipation of fault current or transients on a distribution system*

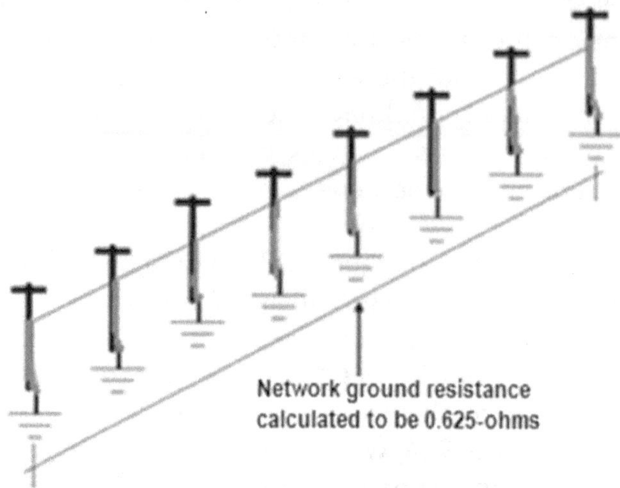

Figure 9.13: *Multiple earth electrodes to improve the dissipation of fault current or transients on a secondary installation*

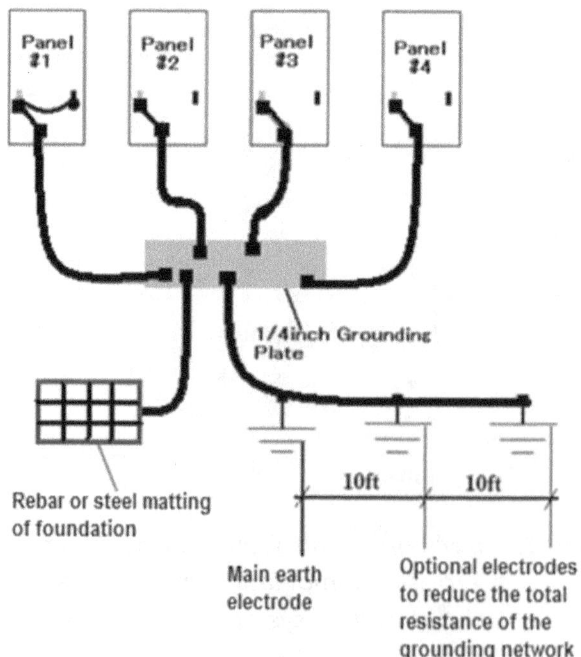

Figure 9.14: *Using more than one earth electrode or earthing method to earth a distribution panel*

9.11 Connecting to the General Mass of Earth

The resistance of the cable connected to the earth electrode does not affect the reading reflected in the instrument when measuring the resistance of the electrode, but it will affect the let-through current and voltage in the event of a fault which is as a result of transients such as lightning.

It is vital to use the correct size for all main earthing conductors so as to prevent blow-off or feedback in the event that high current or voltage should pass through the installation grounding system. Table 9.6 provides the main earthing conductor sizes in accordance with main circuit breaker sizes.

The I^2t on an installation represents the magnitude of short-circuit current produced. However, the damage caused by short-circuit current is *"the current squared multiplied by the fourth power of the time the fault lasts (I^2t^4),"* where I is the fault current incrementing with time t.

Therefore, it is for this reason that earthing arrangements for all electrical installations shall be designed in such a manner so as to prevent the general mass of earth being used as the sole return path for high fault current to the transformer.

Table 9.6: Main earthing conductor sizes in accordance with main circuit breaker sizes.

Circuit Design Current	AWG or Kcmil	mm^2	No. of Rods 8ft
60	8	4	1
80	6	10	1
100	4	16	1
120	3	25	1
140	2	25	1
150	1	35	1
190	1/0	50	1
220	2/0	50	1
250	3/0	70	1
300	4/0	95	1
350	250	120	1
400	300 & 350	150	1
450	400	185	1 – 10ft, ¾
500	500	240	1 – 10ft, ¾
550	600	300	2 – 10ft, ¾
650	750	300	2 – 10ft, ¾
700	1000	500	2 – 10ft, ¾

(Adapted from www.Global-Electron.com August 22, 2004)

9.12 Conclusion

The characteristics of the soil to which a fault current is to be dissipated significantly affects the rate and the magnitude at which the dissipation of current and voltage takes place during a fault or a short circuit. Some soils are classified as corrosive. This corrosive characteristic is sometimes referred to as "aggressive condition" which tend to damage metals which are unable to withstand corrosion. In addition, fault current is not limited to faults which are created on an installation, but also lightning, and other line transients. All faults and use the same designed grounding path for the dissipation of high current or voltage. Consideration must therefore be given to the number of electrodes used to connect an installation to the general mass of earth. This is vital to the reduction of touch and step potential on an installation.

9.13 Test Your knowledge

1. Give three reasons for the difference in soil resistivity throughout an island, even though the soil type is the same.
2. If the general mass of earth is used as the only return path for fault current, how long will it take for a circuit breaker to operate under fault?
3. Explain the expression "let-through current."
4. Draw an illustration of the behavior of fault current through the general mass of earth.
5. The meter pole of a four-pole service is constantly being hit by lightning. What provisions would you put in place to reduce the catastrophic effect of this occurrence?
6. Explain the formula representation of short-circuit current.
7. What would be the value of the let-through current for a clay soil if the installation is located 100 ft from the distribution transformer? The resistivity is 10Ω/m and the voltage at the earth electrode is 120 V.
8. What is the distance that should be maintained between an installation and its earth electrode?
9. Why is it important to maintain the distance between an installation and its earth electrode?
10. What is the minimum size earth electrode which should be used for earthing an installation?
11. State three factors which the dissipation of current to the general mass of earth is dependent on.
12. What is the acceptable network earth reading for power lines and communication facilities?
13. Why are these readings important?
14. Explain the relationship between soils and moisture.
15. What is meant by aggressive soils?
16. What effect does salt in soils have on soil resistivity?
17. Name three earth electrodes besides an earth electrode.
18. What effect does heat have on the resistivity of a soil?

19. Why is soil resistivity considered essential in electrical installation?
20. Name an electrode other than copper which will not be affected by aggressive soils.
21. What is the likely effect of the line side of a circuit breaker being short-circuited?
22. List three soils with minimum aggression which could be used to fill an earth pit.
23. Give three reasons why galvanized earth electrodes are not recommended for use in soils which are highly corrosive.
24. Choosing the lowest soil resistivity in Table 9.1, calculate the maximum let-through current of that soil for an installation located 20 ft from the utility transformer.

Bibliography

i. Avallone, E. A. and Baumeister, III T, 1986, Marks' *Standard Handbook for Mechanical Engineers*. The Ninth ed. Pages 6-62

ii. The Institute of Electrical Engineers—IEE. 2008. Requirements for Electricians. Wiring Regulations. Seventeenth ed. Pages 30-34

iii. Global-Electron.com. 22 August 2004. Wire Sizes. Retrieved on July 8, 2013, from http://www.global-electron.com/wiresizes.htm

iv. The Institute of Electrical Engineers—IEE. 2002. *On-site Guide*. Sixteenth ed. Page 11

v. Schneider Electric. June 2003. *HV Training Manual*. Pages 7-12

vi. The Institute of Electrical Engineers—IEE. 2008. Requirements for Electricians. Wiring Regulations. Sixteenth ed. Pages 46-47.

vii. Lightning and Surge Technologies. Earthing Techniques. Retrieved on July 8, 2013, from http://www.lightningman.com.au/Earthing.pdf

viii. Lyncole XIT Grounding. Soil Resistivity, Testing Four Point Wenner Method. Retrieved on June 14, 2012, from http://www.lyncole.com/uploads/Lyncole_Ground_Test_Methods.pdf

ix. James, M. R. and Arthur C. August 1994. Conditions Contributing to Underground Copper Corrosion. *American Water Works Association Journal*. Retrieved on April 3, 2011, from www.copper.org/resources/properties/protection/underground.html

x. IEEE Std 142-1991, IEE Recommended Practices for Grounding of Industrial and Commercial Power Systems. Retrieved on June 14, 2012, from http://hmin.tripod.com/als/ccs/docs/pdf/Greenbook.pdf

xi. Metal Gems. Copper Earthing Systems. Retrieved on April 10, 2012, from www.mehta-group.com/copper-earthing-systems.html

xii. Mike Holt Enterprises. Opening Circuit Overcurrent Protection Device to Clear Line-to-ground Fault. Retrieved on October 10, 2011, from

xiii. Http://www.mikeholt.com/mojonewsarchive/All-HTML/HTML/GFCI-Receptacles-Without-Ground~19991230.php

xiv. Piantini, A. and Gualberto, L. Lightning Protection of Overhead Power Distribution Lines. Retrieved on September 21, 2011, from http://ebookbrowse.com/invited-lecture-4-pdf-d41833750

xv. Russell, M. J. The Impact of Mains Impedance on Power Quality. Retrieved on November 2, 2010, from Http://www.powerlines.com/pq2kdoc.pdf

xvi. Hennemann, G. R, Mantel, S. 1995. Soil Briefing Jamaica 1. Retrieved on March 26, 2014. Google: Soil Brief Jamaica 1—ISRIC

CHAPTER 10

Open Neutral Conditions

10.0 Introduction

An open neutral condition (ONC) is the separation of the main system neutral conductor from its source. The effect ONCs have on appliances can be extensive and catastrophic. Furthermore, the variation of current and voltage in the appliances while in operation during an open neutral fault could cause significant damage to sensitive appliances.

10.1 Causes of ONCs and Where They Occur

An ONC condition may result from a loose neutral connection or corrosion. This condition may be due to one or a combination of the following:

1. Corrosion at the transformer terminal
2. Corrosion at the meter facility or meter center
3. Corrosion at the distribution center
4. Burnt-off neutral as a result of a short-circuit compounded with loose connections
5. Broken neutral on service laterals at the point of contact with trees or shrubs as shown in Figure 10.1.

○ **Reference for possible O.N.C**

Figure 10.1: *ONC due to shrubberies or environmental conditions*

Neutral conductors are located at any point in an installation where single-phase voltage is required. Neutral conductors can be located at, but are not limited to, the following points of an installation.

1. Supply transformers
2. Distribution panels
3. Sub-panels
4. Inside a junction box
5. Lighting or outlet points

In addition to the main neutral source, an ONC could occur at any point between the distribution transformer and the consumer's installation.

A single distribution transformer supplying several consumers could develop ONC at its terminals and cause significant damage to appliances for consumers on a wide scale. If in the event a thorough testing and investigation is done, and an ONC is not found and the symptom of the problem still exists, it is highly likely that the fault is located at the transformer terminals. In this situation, the utility company should be called to carry out further investigations and have the problem rectified.

10.2 Effects of ONC

HVs or LVs resulting from ONCs can cause severe damage to the equipment. In a residential situation, during an ONC, if a blender is in use while a television and the lights are on, significant distortion lines or erratic behavior on the TV screen and the narrowing of the TV screen as shown in Figure 10.2 will occur. In addition, one may also observe some lights become bright while others become dull. See Figure 10.3.

Figure 10.2: *Erratic behavior of TV1 and TV2*

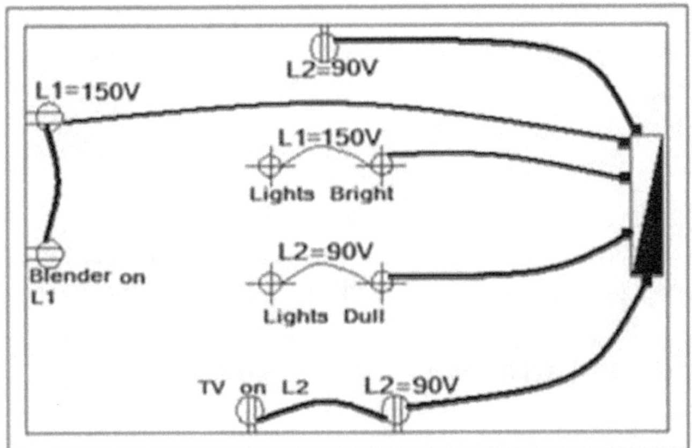

Figure 10.3: *Typical variation in open neutral voltage*

10.2.1 Effects of ONC on Metal Frames of Appliances

An ONC at the utility's transformer during a short circuit on the consumer's installation will result in conductive parts (e.g. metal appliances—refrigerator, metal panel covers, and such accessories) on that installation becoming live. Hence, varying and dangerous touch voltage could be present on appliances and devices. The earth electrode will reduce the effect of the touch voltage. However, an earth electrode will not remove dangerous voltage from the installation. As a result, electrical shock and electrocution are highly possible.

In addition to metal frames, "all conductive parts on an installation must be connected in a manner which creates a permanent low-impedance path from all points of an installation to the electrical supply source."

"Source" in the preceding statement is making reference to a distribution panel, and a low-impedance path is created by connecting all conductive or metal parts of an installation to the neutral or the grounding terminal of a distribution panel. A cable insulated with green insulation signifies grounded conductor. This color-coded conductor must always be used for circuit grounding and bonding of equipment.

It is imperative that all metal frames are connected to one point of an installation, namely the panel, to shunt dangerous potential from the metal frames during a fault to the general mass of earth.

10.3 Open Neutral and Circuit Breakers

Circuit breakers will not protect an electrical installation that develops a short circuit during an ONC. ONCs are created at the utility's transformer or at any point between the transformer and the distribution panel. The automatic operation of circuit breakers will become active only if the open neutral on the installation is corrected. See Figure 10.4.

Figure 10.4: *ONC due to shrubberies or environmental conditions*

10.4 Utility Companies and ONC

Utility companies are responsible for correcting line faults which are created at any point between the transformer and the line side of a metering facility. Consumers are responsible for faults which are created at the load side of a metering facility or at any point beyond the load side of the installation meter facility. After corrections are made, the low-impedance path of the installation or circuit via the neutral conductor to the distribution transformer will be restored, and the circuit breakers will operate under fault conditions.

10.5 Earth Resistance Readings and ONC

An earth electrode with a resistance value between 1-200 Ω will not protect an installation against touch voltage and current. Voltage will be present on the installation at whatever value the earth resistance may be if the circuit breaker does not operate during fault current.

The earth electrode is a means of connecting the installation to the mass of earth for current dissipation during a fault. This connection reduces the effective touch voltage which would have a significant effect on life and property. It is for this reason that the

NEC provides guidance of the resistance of an earth electrode. If a soil condition prevents a specific earth resistance from being met, section 250-56 of the NEC stipulates:

> If a single rod, pipe or plate has a resistance to earth 25 Ω or less, the supplementary electrode shall not be required.

Tables 10.1 and 10.2 developed by Mike Holt, NEC training show the effect earth resistance may not have on appliances during ONC. The ground resistance readings in these tables are based on the three-point test method of an earth electrode (see: www.MikeHolt.com).

In both tables, the voltages in L1 and L2 change to a higher or lower value based on the load demand. The changes in line current are also shown along with the touch voltages and the variation in ground resistance reading (See: https://www.youtube.com/watch?v=Yg6G5VUSsWA).

Table 10.1: Ground resistances varying between 100 Ω to 10 Ω

Line to Neutral Volts	120				
Line 1 Calculated load in Amperes	5				
Line 2 Calculated load in Amperes	15				
Service Neutral Impedance	100	***200***	10000	50000	1000000
Ground Resistance	100	***75***	50	25	10
Neutral/Ground Parallel Resistance	50	54.55	49.75	24.99	10.00
Line 1 - Actual Load Current	7.23	7.25	7.23	15.00	15.00
Line 1 - Load Operates at Volts	173.57	174.05	120.00	120.00	120.00
Line 1 – Over-voltage	53.57	54.05	0.00	0.00	0.00
Line 2 - Actual Load Current	8.3	8.24	15.00	15.00	15.00
Line 2 - Load Operates at Volts	66.43	65.95	120.00	120.00	120.00
Line 2 – Under-Voltage	53.57	54.05	0.00	0.00	0.00
Parallel Neutral/Ground Path in Amperes	1.07	0.99	0.00	0.00	0.00
Touch Voltage from Metal Parts to earth	53.57	54.05	0.00	0.00	0.00

(Adapted from the National Electrical Code Internet Connection, 2002. Also see: www.MikeHolt.com and https://www.youtube.com/watch?v=Yg6G5VUSsWA)

Table 10.2: Ground resistances varying between 8 Ω to 2 Ω

Line to Neutral Volts	120				
Line 1 Calculated load in Amperes	5				
Line 2 Calculated load in Amperes	15				
Service Neutral Impedance	100	200	10000	50000	1000000
Ground Resistance	8	5	4	3	2
Neutral/Ground Parallel Resistance	7.41	4.88	4.00	3	2.00
Line 1 - Actual Load Current	6.38	6.12	6.00	5.83	5.62
Line 1 - Load Operates at Volts	153.15	146.91	143.99	140.00	135.00
Line 1 – Over-voltage or Under Voltage	33.15	26.91	23.99	20.00	15.00
Line 2 - Actual Load Current	10.86	11.64	12.00	12.50	13.13
Line 2 - Load Operates at Volts	86.85	93.09	96.01	100.00	105
Line 2 – Over-voltage or Under Voltage	33.15	26.91	23.99	20.00	15
Parallel Neutral/Ground Path in Amperes	4.48	6.00	6.00	6.67	7.5
Touch Voltage from Metal Parts to earth	148.35	23.99	23.99	20.00	15

(Adapted from the National Electrical Code Internet Connection, 2002. Also see: www.MikeHolt.com and https://www.youtube.com/watch?v=Yg6G5VUSsWA)

The data in Tables 10.1 and 10.2 are based on one earth electrode or rod in different locations having different resistances. It is essential to provide an additional rod or rods as required to lower the connected resistance to the general mass of earth at an installation. The formula $Rt = \frac{1}{\left[\frac{1}{R1} + \frac{1}{R2}\right]}$ could be used as a rule of thumb for calculating the total resistance of rods in parallel. Therefore, if two rods have a resistance of 100 Ω each, then the total resistance of these rods using the formula would be 50 Ω.

10.5.1: Case Study

A homeowner experiences an intermittent electrical shock upon touching the cover of an electrical panel and the metal frame of his electric stove. The current in the earth conductor was determined to be 28 A. Further investigations were carried out, and a 20 A circuit breaker was found with a phase-to-ground short.

An investigation was also carried out to establish why the circuit breaker did not trip as a result of the fault current. It was found that the bonding between the ground and the neutral terminal was missing; this bonding is to create a low-impedance path between the neutral and the ground terminal, which will cause the circuit breaker to operate or trip when a fault is present.

The bonding between the neutral and the earth terminal was done, the fault was cleared, and the circuit breaker was restored to normal operation. A final test was done to determine the ground resistance, and the current reading after the fault was corrected.

A ground resistance of 18 Ω and a current reading of 0 A were found. This indicates that the installation was cleared of all fault currents to ground. The installation was restored to satisfactory working order and will now facilitate the opening of a circuit breaker in the event of a short circuit in the future.

The case study shows, "the earth should not be relied on to provide a substitute neutral path to the transformer if an ONC should occur." In addition, it shows that appliances operating on a defective neutral will suffer some form of damage if it is not corrected in a timely manner. Figure 10.5 shows the earth or the general mass of earth being used as the only return path to the transformer during an ONC.

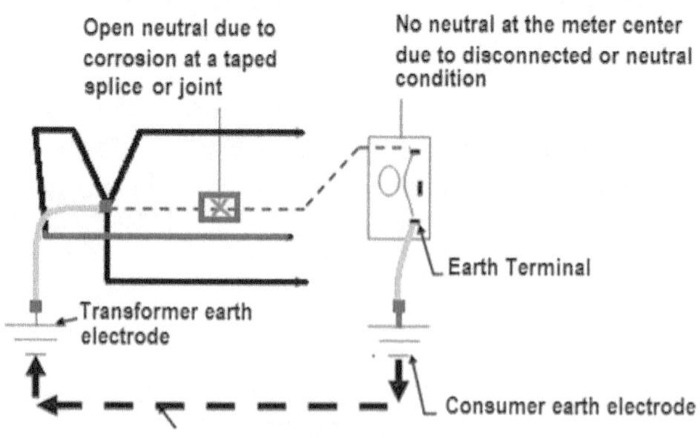

Figure 10.5: *Earth loop (follow dotted lines)*

Earth Loop/Return Path—High Impedance through the general mass of earth is normally above 50 Ω; rarely readings less than 50 Ω will be found.

As it relates to fault current and soil impedance, the impedance of most soils in the Caribbean is usually not less than 50 Ω and most likely, it will be above 100 Ω. This value will be difficult to achieve due to varying geographical conditions and soil types. With this being said, there will be an insufficient flow of fault current if the earth were to be used as the only fault current return path for any installation or equipment. As a result, "less than 1 A for a 120 V line-to-ground fault will flow depending on the soil impedance."

10.6 Conclusion

ONCs are one of the major problems found on electrical installation. Open neutral faults are the most common fault and are sometimes the most difficult to locate. These faults are mainly hidden in joints which crimped or clamped and then taped. If open neutral faults are not corrected immediately, appliances are likely to become damaged by HV over time. It is for this reason that serious consideration must be given to the type of dissimilar material used on power system. Where dissimilar material cannot be avoided, antirust solution should be used on these joints to reduce corrosion.

10. 7 Test Your Knowledge

1. Explain the phrase "open neutral condition."
2. List five areas of an installation where an ONC could be found.
3. Explain the behavior of current and voltage in an installation when there is an ONC.
4. What effect will an earth electrode reading of 5 Ω have on the protection of appliances if there is an ONC?
5. If the main earth conductor of an installation has a current of 30 A, explain in one paragraph what may cause this current to flow. What corrective measures should be taken?
6. List three causes of an ONC.
7. Three earth electrodes have readings 150 Ω, 100 Ω, and 70 Ω respectively. What is the total resistance of the earth electrodes when connected to a common terminal?
8. What effect will the network resistance in question seven have on transients?
9. What effect will an ONC have on the operation of a circuit breaker if there is a fault?
10. What is the maximum fault current which will reach the transformer via the general mass of earth, if a 120 V phase touches a metal frame if the resistivity of the earth is 250 Ω?

Bibliography

i. Donnelly, E. L. 1985. Electrical Installation Theory and Practice. Thomas Nelson and Sons Ltd. Page 5

ii. National Fire Protection Association. 2008. National Electrical Code. National Fire Protection Association. Section 250-56, 250-50, and 250-52

iii. National Fire Protection Association. 2008. National Electrical Code. National Fire Protection Association. Section 250-64

iv. The Institute of Electrical Engineers—IEE. 2002. *On-site Guide,* Sixteenth ed. Pages 78 and 33

v. The Institute of Electrical Engineers—IEE. 2002. *On-site Guide.* Sixteenth ed. Page 79

vi. Mike Holt Enterprises. Danger of Open Neutral Conditions. Retrieved on January 8, 2011, from http://www.mikeholt.com/mojonewsarchive/NEC-HTML/HTML/DangerofOpenServiceNeutral~20020816.htm

CHAPTER 11

Harmonics

11.0 Introduction

An electrical system will generate currents or voltages with frequencies that are *integer or numeric* multiples of the fundamental power frequency. These numeric multiples of the fundamental frequency are referred to as harmonics. Harmonic voltages or currents (sometimes referred to as carbon footprints) generate magnetic waves which may cause interference to nearby communication systems.

11.1 Harmonic Frequencies

Harmonics are currents or voltages with frequencies that are *integer or numeric* multiples of the fundamental power frequency being 50 or 60 Hz (Hertz). For example, if the fundamental power frequency is 60 Hz, then the second harmonic is 120 Hz., the third is 180 Hz, etc. The prolific and increased use of simple household electronics has made the avoidance of harmonics totally impossible. When harmonic frequencies are prevalent and excessive, domestic appliances may malfunction, and electrical power panels including transformers become mechanically resonant due to the magnetic fields generated by the pollution of harmonics. Simply put, buzzing sound for different levels of harmonic frequencies may be heard. This buzzing sound is referred to as humming and can be a nuisance.

Harmonic frequencies ranging from the third to the twenty-fifth are the most common range of frequencies measured in electrical distribution systems. Currents produced with these frequencies in distribution and transmission lines, create a magnetic field which could induce voltage in telephone lines which are located in close proximity to the power lines. Such effect will be noticed in the form of buzzing sounds while the phone is in use. Another significant effect of harmonics is neutral stress. This is a complicated problem in power systems and, if not monitored, could cause significant damage to power systems and appliances.

11.2 Neutral Stress

A neutral conductor in an electrical system is critical in balancing the system for single-phase 120 V appliances and equipment. The neutral is the only viable return path for fault current to the neutral point of the transformer. For this reason, significant attention should be given to neutral conductors when designing an installation or troubleshooting faults.

The main neutral conductor in any electrical installation is considered as the conductor which is most likely to undergo excessive stress from short-circuit current and harmonics. These stresses will cause the neutral conductor to overheat. The main neutral conductor of any installation, whether single phase or three phases, undergoes significant stress; therefore, this conductor must be the same size or larger than the main phase conductor in order to mitigate against overheating. Note, however, that under normal conditions, the neutral current is less than the total current drawn from the supply.

11.3 Linear and Nonlinear Loads

There are two types of loads found on electrical systems; they are linear and nonlinear loads. Linear loads "draw current proportionately with respect to the supply voltage," resulting in sinusoidal or even-current waveforms, whereas nonlinear loads "draw current disproportionately with respect to the supply voltage," resulting in non-sinusoidal current waveforms. Nonlinear loads establish odd harmonics such as third, fifth, and seventh harmonics along with other multiples of harmonics in the neutral conductor. See examples of linear and nonlinear load below loads in Figures 11.1 and 11.2.

11.3.1 Examples of Linear Loads

Loads that are resistive are classified as linear. Examples are:

- Electric stove
- Electric kettles
- Electric iron
- Curling iron
- Toaster

11.3.2 Examples of Nonlinear Loads

Loads that incorporate inductive and capacitive (reactive) elements or features are usually classified as non-linear. Examples are:

- Gas discharge lamps
- Semiconductor power-control devices, (diodes, transistors, silicon control rectifiers (SCRs), triode alternating current switch (TRIACs))
- Computers
- Television and stereo systems
- UPS systems
- Transformers (primary winding magnetizing current is usually non-sinusoidal due to the saturation curve of the core)
- Incandescent lamps (slight asymmetrical currents)
- Hair driers
- Arcing equipment such as welders and arc furnaces

A distribution panel shown in Figure 11.1 supplies loads which are on a balanced 240 V/120 V star distribution system or 220 V/110 V delta distribution system or single-phase (1 ph) 3-wire system. Phases and neutral conductors are sized at 250 mcm or 150 mm^2. Ground conductor size is #6 or 10 mm^2 and will have no unbalanced current acting on the neutral. As a consequence, the size of the neutral could be half the size of the line conductors. Loads that are nonlinear and balanced will have a tremendous amount of harmonic or unbalance current flowing through the neutral conductor of the distribution panel or circuits as shown in Figure 11.2. In this situation, it is imperative that all the conductors are of the same size.

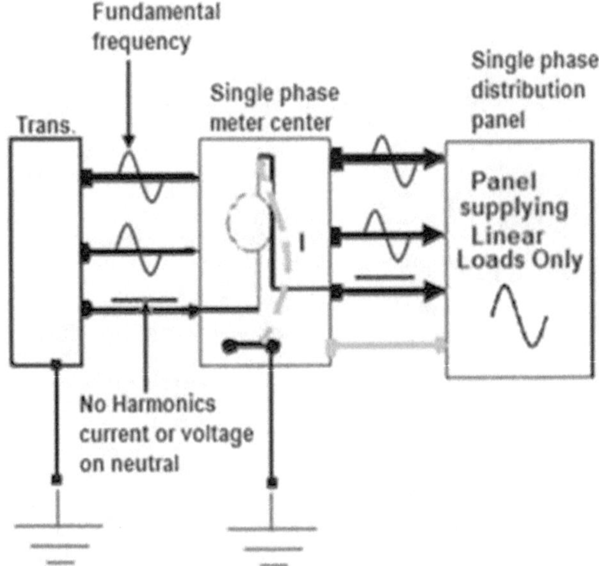

Figure 11.1: *Panel supplying linear loads*

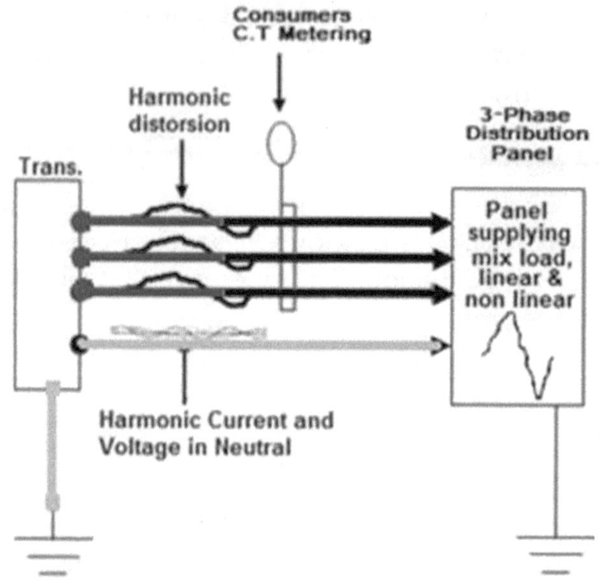

Figure 11.2: *Panel supplying nonlinear loads*

Shown in Figure 11.3 are typical waveforms produced by linear loads.

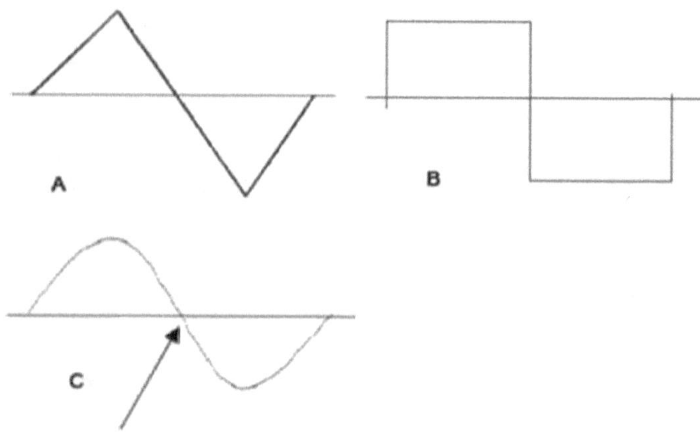

Figure 11.3: *Waveforms produced by linear loads*

Shown in Figure 11.4 are typical waveforms produced by nonlinear loads.

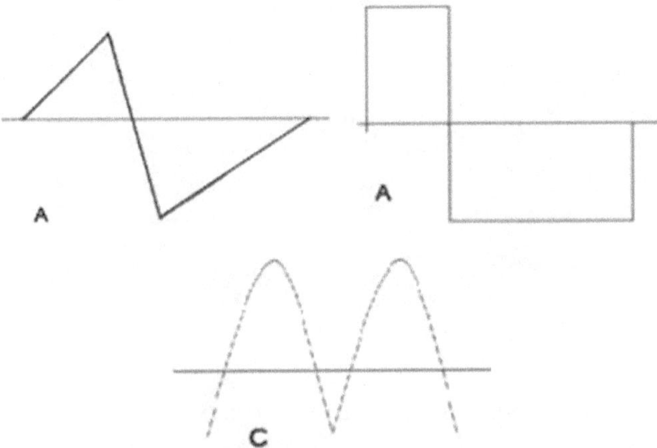

Figure 11.4: *Waveforms produced by nonlinear loads*

11.4 The Effect of Mixed Frequencies on Neutral Conductors

Harmonics in a power system is known as mixed frequency signals and are described as "frequencies that are integer multiples of the fundamental source frequency." Integer multiples representing harmonics such as 3rd, 6th, 9th, 12th, 15th, 18th, 21st, and so on are known as triplen harmonics. These odd integer multiples significantly contribute to the increase in current and voltage acting on the neutral conductor as shown in Figure 11.5. Even-numbered multiples of harmonics such as 6th, 12th, 18th, etc, are not generally significant.

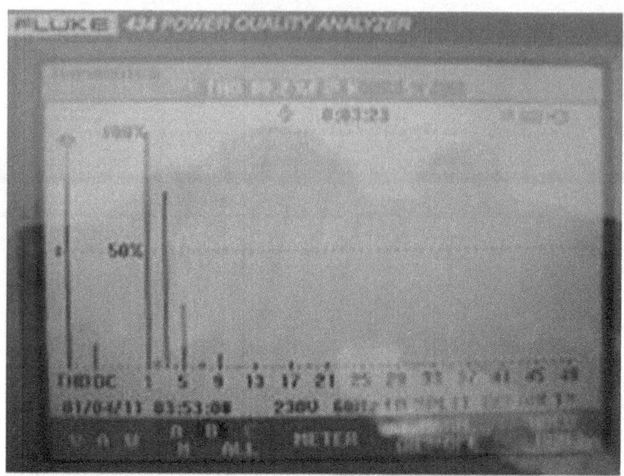

***Figure 11.5**: Triplen harmonics*

In Figure 11.6, increase in the neutral current is due to the adding and circulation of integers or triplen harmonics, via the neutral conductor. This mainly occurs in star or *Y* distribution systems with *Y* connected loads. Equipment returning harmonics current on neutral conductors is not limited to star-connected loads. Single-phase equipment also produces harmonics, returns current, and increases current on neutral conductors. Harmonics do not have 120° phase shifts. It is for this reason that returning harmonics is added in the neutral conductor and will not cancel each other to give a resultant of zero current and voltage. See Figures 11.6 and 11.7.

Third or triplen harmonics produced in delta (Δ) supplies and Δ loads are contained within the source which produces it, thus preventing distorted voltage or additive current occurring in the neutral conductor. Delta supply and delta loads are the ideal solution to problems relating to harmonics, for example, 3-phase motors.

Single-phase nonlinear loads have no means of trapping harmonics; consequently, these loads distribute and receive the same harmonically rich current, injected in the supply phase conductor. Figure 11.6 shows the fundamental waveform and the associated harmonics on the neutral conductor. High harmonic levels or THD above five percent (5%) of the total load current can be extremely harmful to power systems. These added currents may cause the following:

1. Shutdown or malfunctioning of computerized equipment.
2. Increase in wear in motors and other equipment problems for neighboring consumers
3. Failure of light fixtures

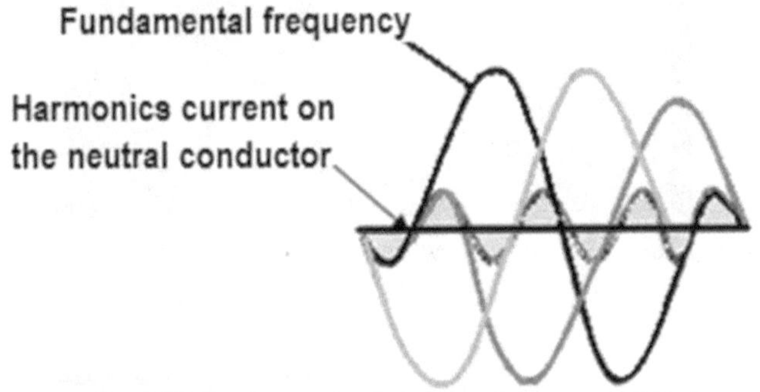

Figure 11.6: *Current due to harmonics on the neutral conductor*

11.5 Total Harmonic Distortion (THD)

Practically, THD is the sum of the total integer multiple of the fundamental waveform. THD does not only affect loads but also supply transformers or any such magnetizing equipment or load. If the THD present on transformers or machines is above 5% of the total load current and greater than 3% for a single appliance, it is possible that overheating of appliances and malfunctioning of sensitive equipment may take place, thus causing deteriorating stress, which would reduce the life of the affected transformer, appliance, or equipment. The effects of THD captured on a FLUKE power quality analyzer are displayed in Figures 11.7, 11.8, and 11.9.

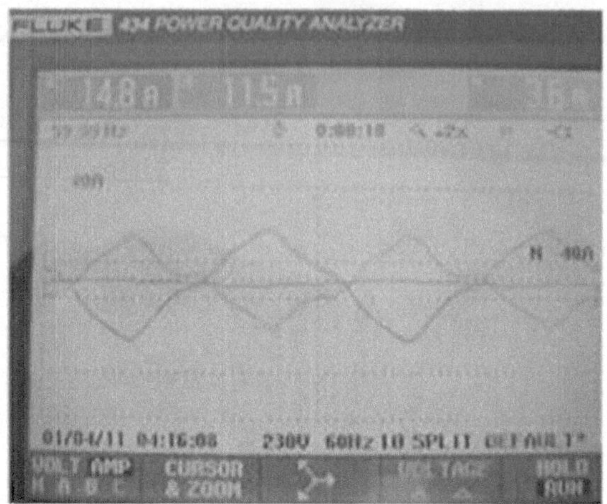

Figure 11.7: *Distorted waveform neutral currents due to harmonics*

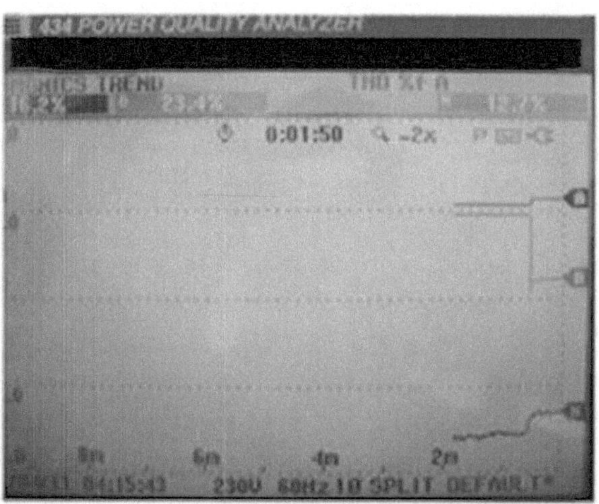

Figure 11.8: *Harmonics trend on black phase, red phase, and neutral*

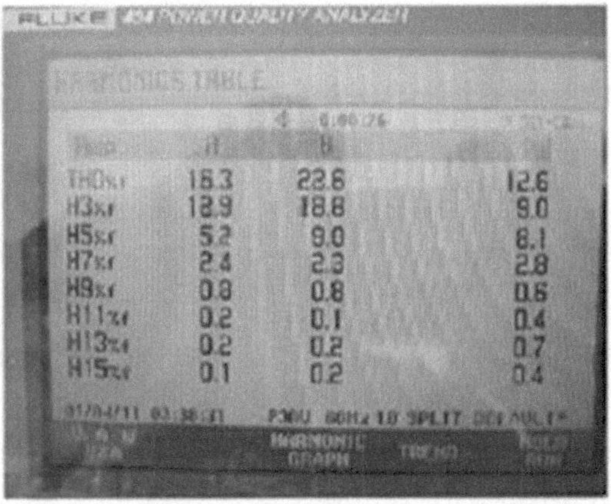

Figure 11.9: *Harmonics table showing % THD and individual harmonics*

Transformers operating at or near full load or at saturation will produce harmonics which will adversely affect the equipment they supply. Under this condition, transformers may cause errors in analogue kWh meters. These errors force the device to rotate between 1.5 to 2% faster. It is for this reason that emphasis should be placed on harmonics pollution. A reduction of the effect noise pollution would also reduce the kWh consumed.

The T1000 high-efficiency harmonic canceling transformer manufactured by Powersmith is highly recommended. This transformer can be used to reduce the effect of generating harmonics and increase power efficiency. For more information about this transformer, refer to www.powersmiths.com.

A distribution system with its entire load perfectly balanced across all phases as well as having all loads linear is referred to as an ideal distribution system. In this system, there is no harmonic pollution and the resultant current is zero in the neutral conductor. The resultant waveform in this installation is perfect fundamental waveform as shown in Figure 11.10. This condition is said to be impossible to obtain.

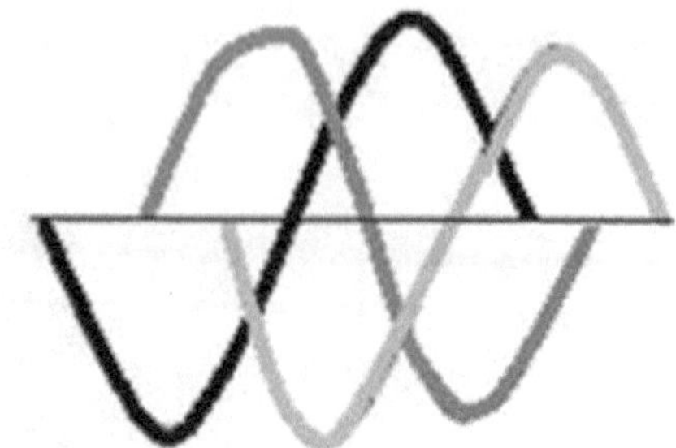

Figure 11.10: *Fundamental waveforms*

In Figure 11.10, the waveforms cancel each other, thereby eliminating all currents in the neutral conductor.

In addition to the perfect balancing theory, the evolution of technological advances brings about a flood of new and advanced electronic devices and equipment which are not harmonic-friendly and contribute a large amount of harmonics in the neutral conductor.

Harmonic friendly devices include switch-mode power supplies (SMPS), d.c. - d.c., and d.c. – a.c. converters were referred to as choppers in earlier years when SCRs were used. Today, Insulated Gate Bipolar Transistors (IGBTs) and Metal Oxide Semiconductor Field Effect Transistors (MOSFETs) are used in the construction of these devices.

Practically, a "balanced power system" will have no negative or zero sequence components. Uniform waveforms will cancel out themselves and therefore will not cause neutral conductors to overheat. However, if the main appliances in everyday activity generate harmonic waveforms, significant destructive current will flow in the neutral.

All generated harmonics fall within a sequence pattern. Each pattern describes the type of effect the harmonics value will have on appliances used on the distribution system as shown in Table 11.1.

Harmonic numbers indicate harmonic frequencies: the first harmonic is the fundamental frequency (60 or 50 Hz). The second harmonic is the component with two times (2 x 60 or 2 x 50) the fundamental frequency (120 or 100 Hz). The third harmonic is the component with three times (3 x 60 or 3 x 50) the fundamental frequency (180 or 150 Hz), and so on. Furthermore, harmonic sequences are grouped into positive (+), zero (0), or negative (-). Table 11.1 provides an overview.

11.6 Harmonics in Induction Motors

According to Sankaran (1999) in the *Fluke User Manual of 1, 04/2007:*

> Positive sequence harmonics in a motor force it to accelerate faster than the fundamental frequency. These harmonics (1, 4, 7, 10, etc.) produce magnetic fields and current rotating in the same direction as the fundamental harmonics.

The *Fluke User Manual* further stipulates that:

> The presence of negative sequence harmonics in motor forces it to accelerate slower than the fundamental frequency. These harmonics (2, 5, 8, 11, 14, etc.) develop magnetic fields and current that rotates in the direction opposite to the fundamental frequency.

And that:

> Zero sequence harmonics (3, 6, 9, 15, 21, etc.) do not produce usable torque, they produce additional losses in machines and causes an increase in the neutral current.

Zero sequence currents produced in a distribution system are also classified as circulating currents; these currents are fault currents which may be due to an unbalanced system (Sankaran, 1999).

In the first two cases described, the motor loses torque; however, in all three cases, heat is produced, causing the motor's temperature to increase. Furthermore, harmonics will disappear if the waveforms are symmetrical, that is, both the positive and negative segments are identical, except for their opposite polarity.

Table 11.1: Harmonic sequence in motors

Order	1st	2nd	3rd	4th	5th	6th
Frequency	60Hz	120Hz	180Hz	240Hz	300Hz	360Hz
	50Hz	100Hz	150Hz	200Hz	250Hz	300Hz
Sequence	+	-	0	+	-	0
Order	7th	8th	9th	10th	11th	...
Frequency	420Hz	480Hz	540Hz	600Hz	660Hz	...
	350Hz	400Hz	450Hz	500Hz	550Hz	...
Sequence	+	-	0	+	-	...

(Adapted from *Fluke User Manual,* April 1, 2007)

In order to reduce the effect of harmonic sequencing in motors, especially for three-phase equipment or appliances, the most appropriate wiring configuration to be employed is the delta as shown in Figure 11.11.

11.7 Torsional Oscillation of Motor Shafts

Table 11.1 confirms the practicality of Newton's third Law. This law states, "When two particles exert forces on each other, the forces are equal in magnitude, opposite in direction and collinear." This means that when the fundamental frequency changes from its original state, something else is directly affected.

Furthermore, changes in the state of the fundamental frequency will result in an interaction between the positive and negative sequence of the magnetic fields and current, thus producing torsional oscillation of motor shafts. Oscillation of motor shafts will result in vibration thereafter, resulting in overheating, expansion of moving parts, and severe damage to the motor shaft and bearings (*Fluke User Manual,* April 1, 2007).

Acknowledging the damage that harmonics can create, it becomes crucial for large industrial plants, factories, and processing plants to implement the use of variable frequency drives (VFDs) for motor application to conduct harmonic evaluation, so as to determine the harmonic distortion levels present on a motor's distribution panels. This is an attempt to reduce the effect of harmonics on the entire distribution system and motors that could be severely affected Figure 11.12 shows the grounding configuration and design method of eliminating stray harmonics (*Fluke User Manual,* April 1, 2007).

Severe torsional oscillation is only significant if a three-phase motor is connected in star or *Y* configuration and is supplied by a star or *Y* configured transformer. In Figure 11.13 the delta connected supply and load which has harmonics, torsional oscillation is significantly reduced. There is no neutral coupled to the supply transformer winding.

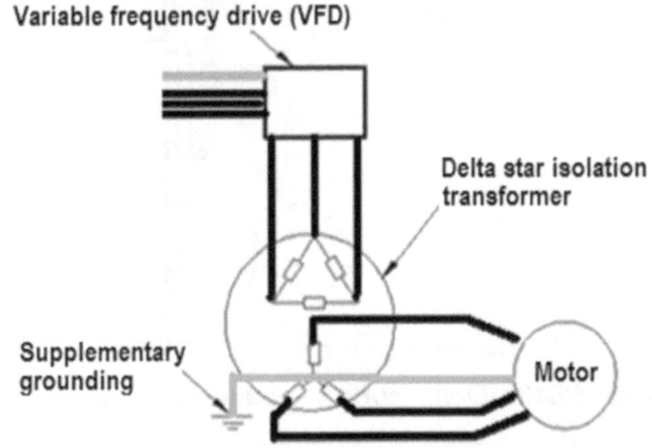

Figure 11.11: *Configuring a VFD to eliminate the effects of harmonic current*

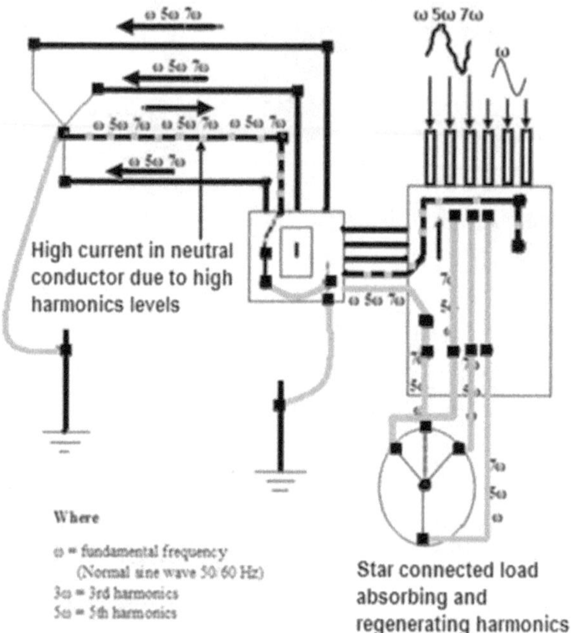

Figure 11.12: *Harmonics polluted system causing neutral stress and torsional oscillation of motor shafts*

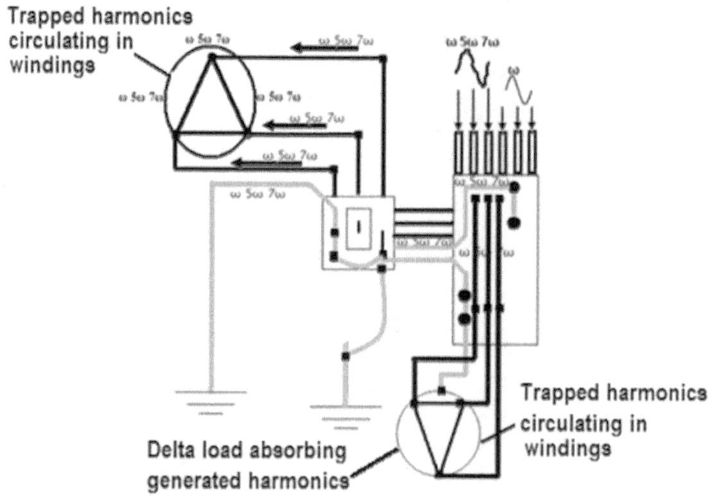

Figure 11.13: *Delta supply and a delta connect load containing harmonics.*

11.8 The Implications of Harmonics

In most domestic and commercial installations, the presence of third harmonic levels poses little or no threat to some appliances. However, motor driven appliances such as blenders can be severely affected. These appliances may overheat and burn out quickly. Additionally, light fixtures or bulbs in circuits which are subjected to severely polluted harmonics will be severely affected. In some cases, the light fixtures may burn out quickly and light bulbs easily blown. These harmonic effects are due to a polluted neutral. This type of neutral pollution is referred to as intermittent pollution and only occurs when the distribution system is severely polluted.

Other possible harmonic problems include the following:

1. Excessive heating and failure of capacitors, capacitor fuses, transformers, single-phase motors, fluorescent lighting ballasts (magnetic), etc.
2. Nuisance tripping of ground fault circuit breakers due to high levels of noise associated with equipment that generate harmonics
3. Heating up of the neutral conductor
4. Noise from harmonics which lead to erroneous operation of control system components
5. Damage to sensitive equipment
6. Communication interference

Utility companies transmit the fundamental waveform 50 Hz or 60 Hz, which is free from harmonics distortions. However, even though the fundamental frequency is on the primary side of the transformer, consumers will experience significant harmonic distortion as each consumer contributes to the THD on a distribution system.

A transformer supplying a small individual consumer will experience far less harmonic levels, depending on the number of electronic devices being used on that installation.

The presence of harmonic distortion can be easily detected on an installation. Figure 11.14 shows the setup of a fluke analyzer to measure harmonics distortion and other values on a three-phase distribution system.

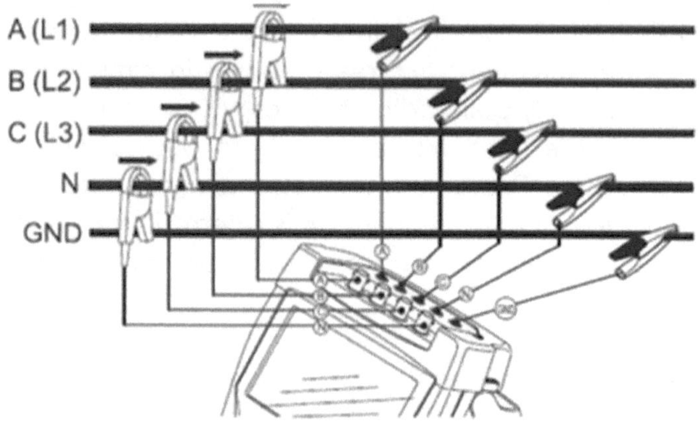

(Adapted from *Fluke User Manual,* April 1, 2007)
Figure 11.14: *How to connect a fluke power analyzer*

Harmonics on an installation can be developed by the electronic devices which are constantly being used on an installation, for example, TVs, stereos, computers, adapters, compact fluorescent lighting fixtures, and any other electronic device or equipment which influence the presence of high- or low-level harmonics. With this information, you will be better able to resolve unusual phenomena such as shutdown, blown bulbs, nuisance tripping, and noise along with many other strange phenomena that may occur on an electrical installation. This recurring phenomena on domestic and commercial installations can be resolved by creating an extremely low impedance path to the general mass of earth. This low-impedance path will act as a shunt, thus diverting harmful harmonic current to the earth.

11.9 Conclusion

Harmonics and harmonics current are sometimes referred to as carbon footprints. Harmonics current significantly affects industrial machines and motors where huge amount of electronics are used as a part of normal operations. Equipment such as VFDs and other such loads produce nonlinear waveforms. Waveforms produced by harmonics significantly affect motor shafts and will produce significant heat which will affect the operation of machineries.

11.10 Test Your Knowledge

1. Explain the effect linear and nonlinear loads have on the main neutral conductor in an installation.
2. Sketch two waveforms which represent linear and nonlinear loads.
3. What is harmonics?
4. What are the values of fundamental frequency?
5. What is the name given to the integer which will significantly affect the current in a neutral conductor?
6. Draw and explain the configuration which will significantly control the transfer of harmonics from a three-phase motor.
7. Draw and explain a diagram of a distribution system which allows the transfer of harmonics in a neutral conductor.
8. Draw a diagram to demonstrate the fundamental waveform of a three-phase distribution system and the effect of 3rd harmonics on the neutral conductor.
9. What is meant by THD?
10. What is the maximum THD allowed on any power system?
11. Why is it important to balance all loads evenly across a distribution panel?
12. Describe the effect of negative, zero, and positive harmonic sequences in a motor.
13. State six possible harmonic problems which may occur on a distribution system as a result of harmonics.
14. Name one instrument which can be used to measure harmonics.
15. Draw and explain the setup of an instrument for measuring harmonics.
16. Give one reason for the insignificance of even harmonics.

17. List three causes for the presence of harmonics in a distribution system.
18. Define what is meant by PFC.
19. Define the term *impedance*.
20. Explain what is meant by current downstream and current upstream.
21. Describe two methods of identifying harmonics on an installation.
22. What is the term used to describe harmonics appliance on an installation?
23. Describe a simple method of correcting harmonics on a domestic installation.
24. If the frequency of a distribution system is 60 Hz, what is the 7th harmonics on this distribution system?
25. What is torsional oscillation?
26. What effect, if any, does torsional oscillation have on a three-phase motor?
27. What is the effect of zero sequence harmonics on an installation?

Bibliography

i. Sankaran, C. October 1, 1999. *Fluke User Manual.* April 1, 2007, Harmonics. Chapter 10, Page 305

ii. Sankaran, C. October 1, 1999. Effects of Harmonics on Power Systems—Part 1. 12:00 p.m. http://ecmweb.com/mag/electric_effects_harmonics_power_2/#comment-5116214

iii. Harmonics Distortion. www.galco.com/circuit/PFCC_har.htm

iv. Hearn, S. D., n.d. Basic Understanding of Harmonics in Electrical Systems. http://www.hearneng.com/WhitePapers/Understanding%20of%20Electrical%20Harmonics.pdf

v. Lessons Ins Electric Circuits—Volume 2, Chapter 9. Linear and Nonlinear Loads. http://www.metalwebnews.org/ftp/ac-theory.pdf

vi. The IEE Wiring Regulation, Sixteenth edition states. Wiring System; Part 2, Page 27-29

vii. Power Factor Correction Equipment Harmonic Filters—PQE http://www.pqeltd.com/harmonic-filters.htm

viii. Grady, W. M. and Snatos S. Understanding Power Systems and Harmonics.W. Mack Grady & Surya Snatos, The University of Texas at Austin Austin, TX & Electrotek Concepts, Inc Knoxville, TN respectively. http://patricioconcha.ubb.cl/comprendiendo_armonicos/Understanding_Harmonics.pdf

ix. Total Harmonic Distortion (THD): A Lesson for Lighting Harmony. http://www.geappliances.com/email/lighting/specifier/downloads/Total_Harmonic_Distortion.pdf

x. Ramos A. Jr. January 25, 1999. Treating Harmonics in Electrical Distribution Systems by victor. http://services.eng.uts.edu.au/~venkat/pe_html/pe07_nc8.htm

xi. Robert, E. F. Save Energy and Improve Power Quality! A Case Study—75 KVA Transformer Harmonics. http://www.powerstudies.com/articles/SaveEnergy.pdf

CHAPTER 12

Lightning Protection

12.0 Introduction

Lightning protection refers to measures taken or implemented to preserve life and property from high discharged destructive current produced by an atmospheric phenomenon called lightning.

12.1 Lightning

Lightning is the release of electrical energy created by the accumulation of positive and negative charges in the clouds as shown in Figure 12.1. This is evident in the form of a spark or flash during a thunderstorm.

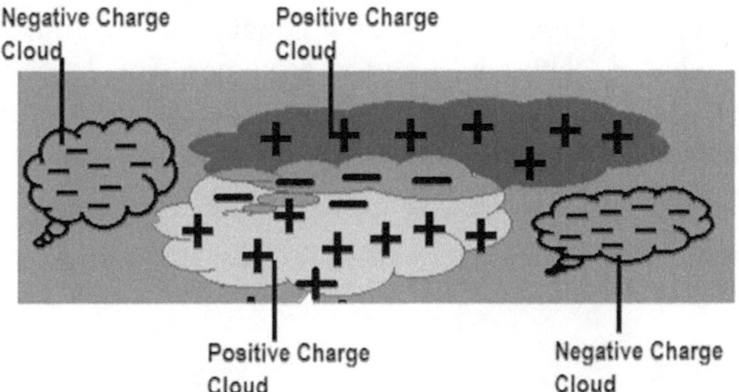

Figure 12.1: *Positive and negative charges produce lightning*

When the temperature in the clouds is below 0°C, it produces ice. Ice particles inside a cloud separate from the cloud as super cold water or ice droplets which are called hailstones. The separation of these particles leaves static charges which when released creates lightning. This basic principle of the production of lightning is thought to be a key element in the development of lightning strokes and leaders as shown in Figure 12.2.

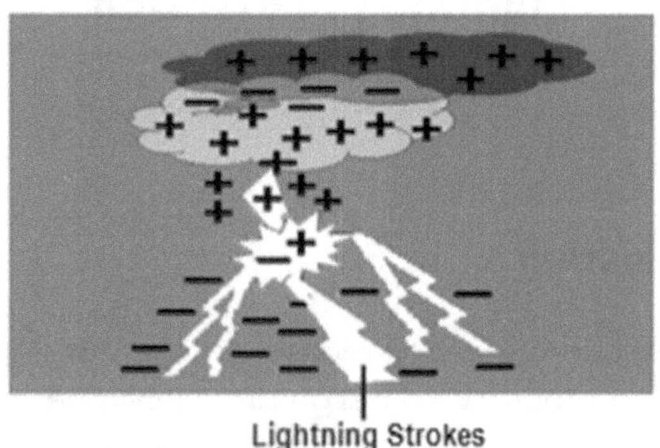

Figure 12.2: *Lightning strokes*

12.1.1 Point of Attraction

Lightning is always attracted to the highest point of a building or the highest structure among a number of buildings. The same is true for trees or power lines above buildings. As shown in Figure 12.3, the points of contact by lightning such as structures, trees, and power lines, all provide a signature of resistance which initiates what is called an upward streamer. The location of a streamer and a stroke is dependent on the cloud which has the most static charges; these clouds are called lightning clouds.

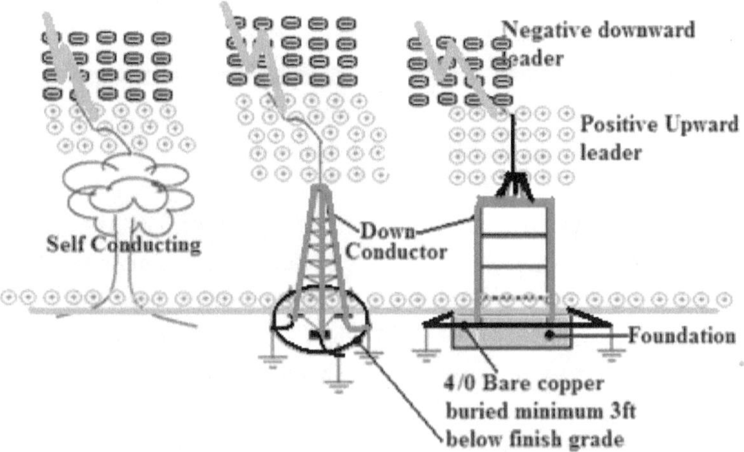

Figure 12.3: *Upward leader from various structures*

Step leader is a charge in the form of multiple steps discharged from the clouds. The field strength of the charge intensifies as it gets close to or within 40 m to 50 m of the earth, creating an upward leader. The upward leader as shown in Figure 12.4 has a positive charge and completes the path to the earth from the step leader which has a negative charge.

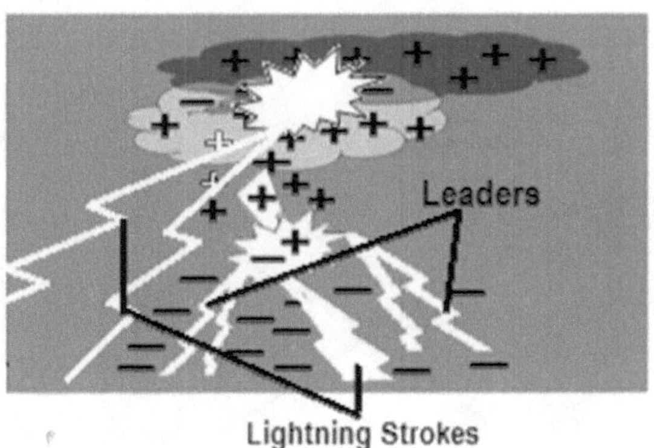

Figure 12.4: *Lightning strokes and leaders*

12.2 Lightning Protection System (LPS)

It is now evident that along with a lightning stroke, there are leaders which can cause significant damage to life and property.

To provide protection against leaders, a series of lightning arrestors or spikes must be installed on the top of high buildings, steel towers, and utility poles. Additional arrestors will capture lightning leaders and strokes, preventing structures from becoming the path for lightning discharges.

Figure 12.5 shows a typical lightening arrestor system. The installation of arrestors, multiple spikes, and bonding of down conductors should be done in accordance with the international building code and the NEC. Installing lightning arrestor systems incorrectly will pose a higher risk of damage to structures and other sensitive equipment. Where multiple arrestors are installed to capture lightning flashover or the leaders, down conductors must be installed in a manner which will prevent flashovers to equipment or apparatus.

Damage from a shoddy arrestor installation could result in fires which lead to structural damage such as foundation fractures. Note that the total current produced by a lightning charge ranges between 3,000 A and 250,000 A; therefore, it is crucial to have as many rods or electrodes connected as a network to the lightning arrestor to assist in quick dissipation of current and voltage which will prevent flashovers and damage to structural foundations.

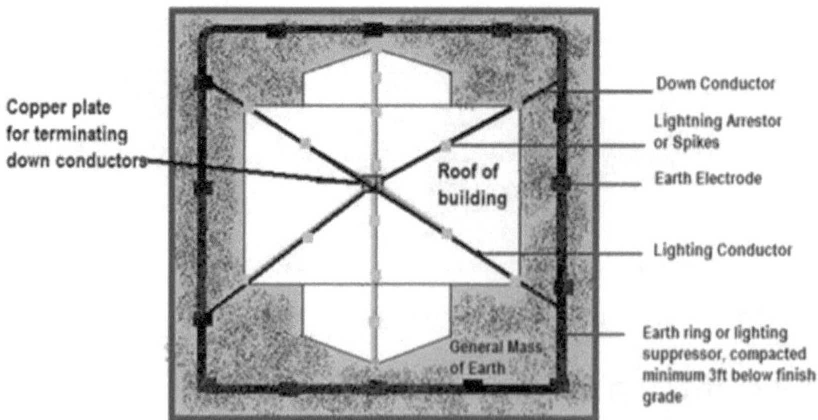

Figure 12.5: *Typical lightning arrestor system*

12.3 Lightning Flashovers

Lightning flashovers are capable of causing catastrophic damage to life and property. It is of immense importance that a minimum of 6′ be maintained between the down conductor and all accessories such as panels and fixtures. As shown in Figure 12.6, the minimum distance must be maintained between all equipment and earth electrodes. If the minimum requirement cannot be maintained, an equipotential bonding must be placed between the down electrode and the electrical system.

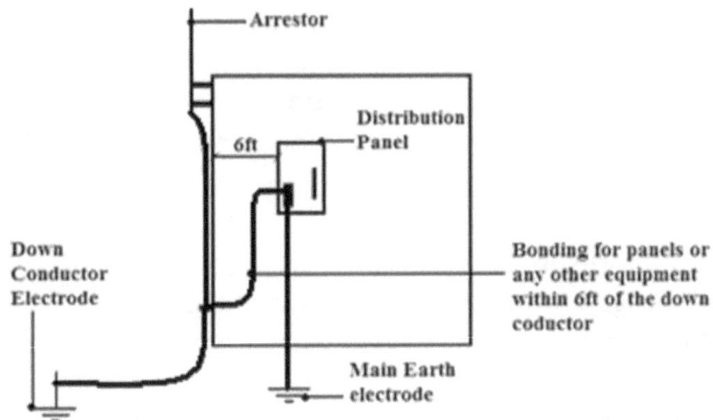

Figure 12.6: *Bonding utilization equipment within 6 feet of LPS*

12.4 LPS Requirements

The LPS network should have a combined resistance to the general mass of earth not exceeding 10 Ω, without taking into consideration any other bonding to other services if necessary. As a rule of thumb, for areas where low resistance is difficult to achieve, the resistance to earth of each down conductor should not exceed 10 times the number of earth electrodes to be installed.

LPS is recommended for all buildings at and above three stories, churches, convention centers, and any such premises.

A building consisting of metal frame working and carrying a LPS must have equipotential bonding done between the main metal verticals and the down electrode.

Service laterals or twisted wire as it is commonly called is a carrier of high surges from lightning. These surges are normally carried by the neutral conductor, even though a reliable lightning system may be a part of a building or structure. This means lightning surges could be generated at any point of a distribution system. This will affect any installation within the affected distribution network. As a result, emphasis must be placed on connectivity to the general mass of earth in order to protect electrical installations from the following:

1. Lightning surges
2. HV surges (transients due to switching)
3. Unintentional contact of service laterals with higher voltage

12.5 Transients through Service Laterals

Lightning strikes and other transients often affect the neutral conductor. In order to reduce the effect of transient through service laterals, supplementary earthing rods are added at the foot of each pole as shown in Figure 12.7. Each electrode will dissipate current and voltage to the general mass of earth. Additional rods or electrodes will reduce the effect that transients would have on the installation.

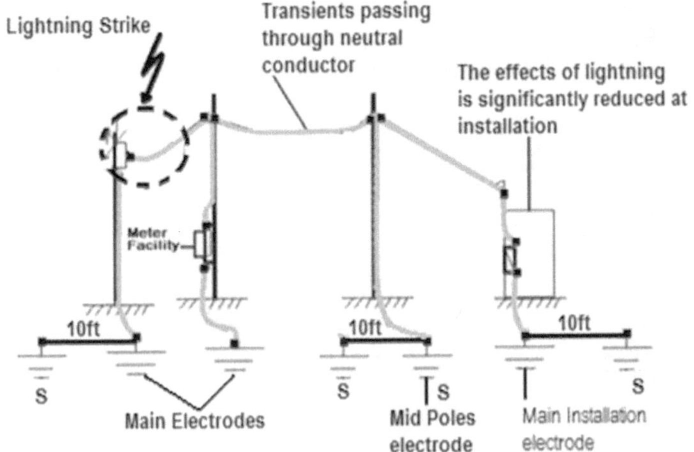

Figure 12.7: *Supplementary earth of service laterals*

According to the IEE *On-site Guide*, domestic line transients are absorbed by the installation of earth electrodes with earth resistances not greater than 200 Ω. However, the NEC states:

> Section 250.53; "If a single rod, pipe or plate has a resistance to earth 25 Ω or less, the supplementary electrode shall not be required.

Poor electrode connections for a lightning arrestor system located above the surface of the earth will not stand up to a lightning discharge of 250 kA. It is therefore essential that proper connections be established to carry thousands of amperes which could be induced by a lightning strike.

The configuration with additional electrodes buried around a building or used by itself will equalize the potential discharged in the general mass of earth and quickly dissipate harmful currents, thus preventing structural damage and electrocution. The configuration in Figure 12.8 will reflect an earth ring reading of zero (0), which does not reflect the true reading of the earth electrode in the ring. Therefore, loop resistance tests between rods should be done to achieve a base resistance of 5 Ω. This test should be done before the earth ring is connected as shown in Figure 12.9.

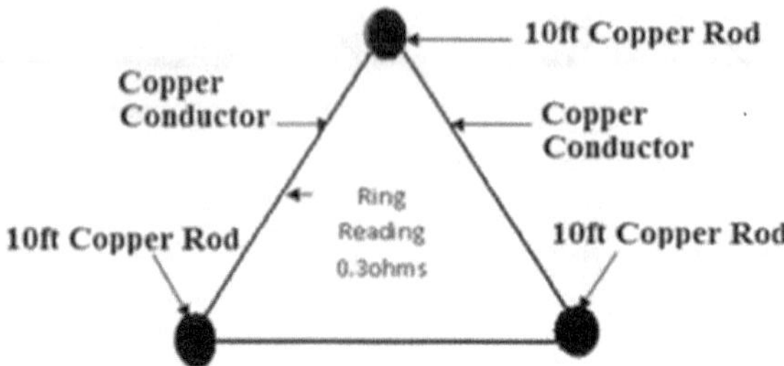

Figure 12.8: *A general earthing arrangement of buried electrodes*

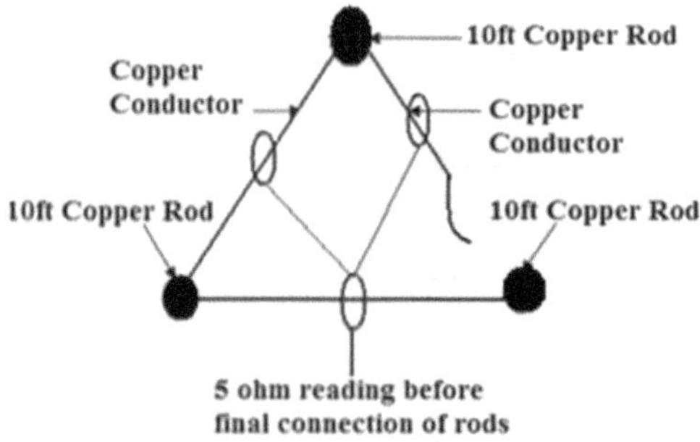

Figure 12.9: *Testing earth electrodes to be connected in delta*

Figures 12.10 and 12.11 show earth electrodes connected in parallel or star configuration. In this earthing arrangement, current and voltage are dissipated to the general mass of earth from the lightning arrestors. The calculation following Figure 12.11 shows the total resistance of a network of electrodes in parallel or what is before a ring configuration is connected. This earthing arrangement is the recommended practice for earth electrodes above the surface or where below surface configurations cannot be met.

The configuration in Figure 12.8 shall be buried at a minimum of 3 ft to the top of the electrodes and shall be of copper rod and bare copper cable or copper rod or copper tape. Each electrode shall have a resistance reading of 25 Ω between the arrestors and the electrode for single-level domestic installations, and 1 Ω-5 Ω for tower utility poles and wind turbine towers.

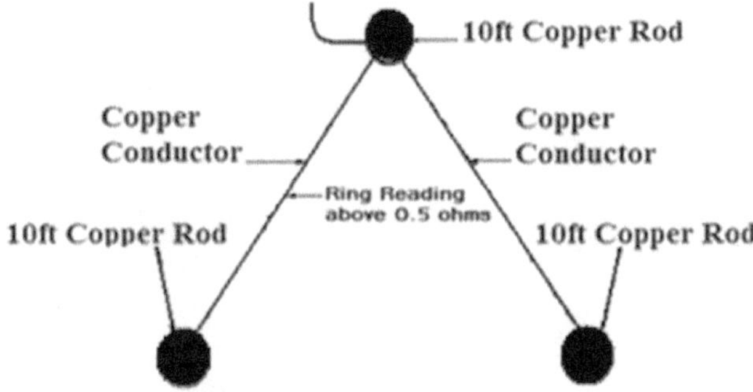

Figure 12.10: *Connecting electrodes in parallel above the surface of the general mass of earth*

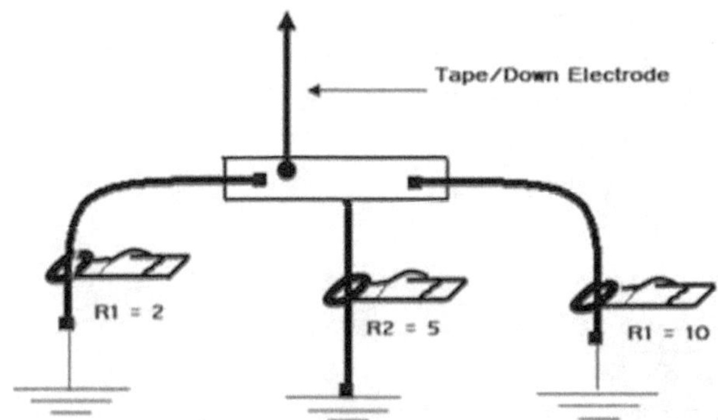

Figure 12.11: *Resistance reading of earth electrodes connected in parallel*

$$\textit{Resistance of earth network} = \frac{1}{1/R1+1/R2+1/R3}$$

$$= \frac{1}{1/2+1/5+1/10}$$

$$= \frac{1}{0.5+0.2+1}$$

$$= \frac{1}{1.7}$$

$$= 0.588\Omega$$

The total resistance of 0.588 Ω means that the current and voltage induced by a lightning strike will have a low impedance traveling to the general mass of earth.

12.6 Lightning Damage

Damage caused by lightning strikes is often blamed on electricity suppliers who do not provide sufficient protection from the effects of lightning strikes. Utility companies will only accept liability where ground resistances of poles supplying the consumer are below standard. Consumers should be aware that damage caused by lightning is not preventable, depending on the level of charge injected in the distribution system. It is for this reason lightning strikes are referred to as acts of God.

Utility companies with excellent customer service will assist their customers with the cost of repairs or replacing damaged equipment when requests are made. This is at the discretion of the supplier and customers must be reminded that utility companies are not obligated to honor such requests.

As a means of precaution, all electrical installations with their service laterals extending more than one pole length must be fitted with a supplementary earth electrode. This supplementary electrode will serve to reduce the effects of lightning transients and other voltage surges which may affect equipment on the installation.

12.7 Protection for Towers

Towers are tall steel structures which are used to mount antennas, transmission lines, wind turbines, stadium lighting, lightning receptors, and a wide range of fixtures that requires elevation from the ground. Tall multi-level buildings designed for residential and commercial purposes could also be used as towers. Heights of towers could range from one hundred feet (100 feet) to more than six hundred feet (600 feet).

Lightning is a tower's greatest enemy. Their protection relies heavily on an exceptionally solid grounding system. The following must be considered when designing LPSs for towers and tall buildings:

1. Observe and follow engineering designs carefully.
2. Install large down conductors.
3. Use multiple down conductors.
4. Do not create a loop or ring at the top of zinc roofs or on the top of wooden buildings.
5. Create an earth ring configuration around foundations.
6. Install several earth rods in the earth ring.
7. Connect earth rings to the rebars of the tower foundation.
8. All joints must be cad welded.

Figures 12.12 and 12.13 are typical designs for grounding of towers. These designs are not limited to towers only and could be used for other similar protection where required.

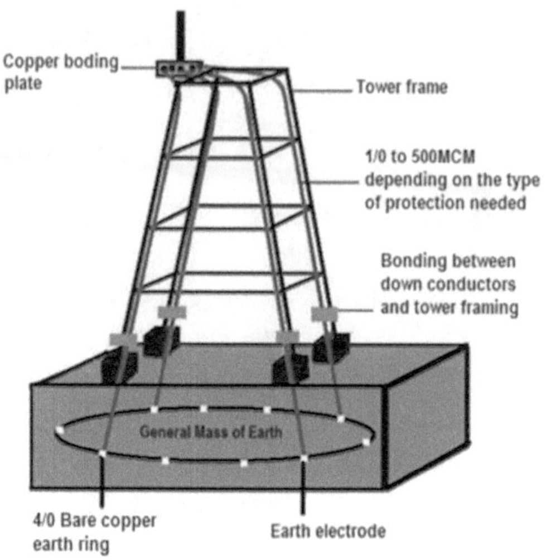

Figure 12.12: *Typical tower with LPS*

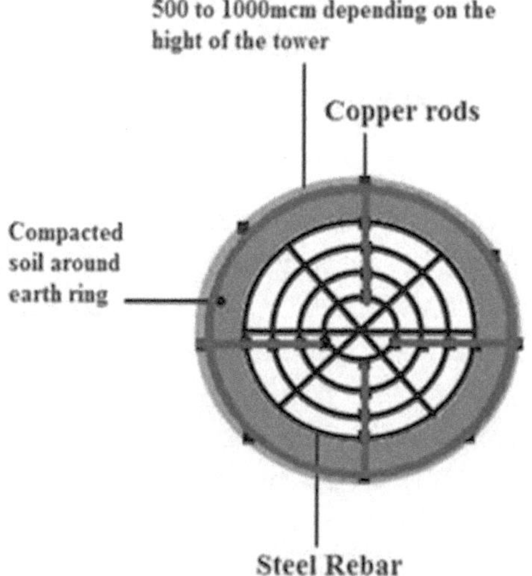

Figure 12.13: *Typical earth ring for tower protection*

12.8 Conclusion

The study of lightning and the path which lightning takes to release its energy must be fully understood before designing the intended path for lightning discharge. High structures and towers or poles are easy targets for lightning. Arrestors or lightning distracters must be carefully designed for structural lightning targets.

12.9 Test Your Knowledge

1. Why is it necessary to maintain 6 ft between the down conductor of a lightning arrestor and the power distribution system?
2. Draw a sketch showing what is to be done if the minimum distance between the distribution system and the LPS is not maintained.
3. Why is it important to attach supplementary earth electrodes to an overhead service cable?
4. What is the best method of protecting an installation from lighting?
5. What type of installation requires a LPS?
6. Draw and label a LPS for a tower.
7. Explain the importance of the delta configuration to earth electrodes on a LPS.
8. What is the main difference between the delta-connected earth electrodes and the star-connected electrodes?
9. Show in detail how you would protect an installation with three service poles from lightning transients.
10. List and explain two natural occurrences in the electrical industry.
11. What is the range of current which could be discharged by one lightning strike?
12. Why are ring connections not recommended for roofs and towers?
13. What is the average resistance between electrodes of a LPS?
14. What are the disadvantages of using undersized down conductors?

Bibliography

i. Erico International Corporation. 2013. Wind Power Solution. Retrieved in May 2011, from http://www.erico.com/static.asp?id=38&gclid=CI6dhdWW2a8CFcqA7Qod_yLKCA

ii. Erico International Corporation. 2013. Foundation Grounding and Construction. Retrieved in December 2013, from http://www.erico.com/static.asp?id=43

iii. LM Wind Power. Strike the Lightning Away. n.d. Retrieved in December 2011, from http://www.lmwindpower.com/Rotor-Blades/Products/Features/Add-Ons/Lightning-Protection

iv. Lightning and Surge Protection of Multi-mega Watt wind Turbines. n.d. Retrieved in December 2011, from http://alt.dehn.de/pdf/blitzplaner/Chapters/BBP_E_Chapter_09_16.pdf

v. Staszewski, L. Lightning Phenomenon—Introduction and Basic Information to Understand the Power of Nature. n.d. Retrieved in December 2011, from http://eeeic.org/proc/papers/52.pdf

vi. Ringleb, U. n.d. Lightning Protection. Retrieved in December 2011, from http://www.schunk-group.com/sixcms/media.php/1701/10-11-12_lightning-protection.pdf

CHAPTER 13

First Aid Applications

13.0 Introduction

First aid is the administering of basic lifesaving applications to a person in the event of an injury or sudden illness in the workplace. First aid is critical in electrical engineering and the construction industry. On electrical sites, there are many situations when knowledge of and ability to implement first aid procedures are essential. Importantly, first aid application is essential as we consider the fundamentals of fault current and grounding since many injuries are generated as a result of faults on electrical systems.

13.1 Typical Hazards and Injuries

A basic knowledge of the types of injuries or illnesses, first aid techniques, and their associated applications could prevent death or casualties in the workplace. These associated principles should be kept as vital information to be used where accidents occur and to reduce the risk of fatalities, injuries, or sudden illness.

Injuries which are likely to occur in the workplace include laceration, abrasion, burns, sprains/strains, crush injuries, or chemical poisoning. In addition to other types of injuries, there are other common medical conditions which could occur on work sites such as asthma, epilepsy, heart attack, and hypoglycaemia (in a person with diabetes) among other conditions.

Table 13.1 shows some typical hazards, their typical problems, as well as the typical injuries or illnesses associated with them. Injuries are not limited to those mentioned in this table.

Table 13.1: Hazards and their associated injuries or illnesses

Hazards	Typical Problems	Typical Injury/Illness Requiring First Aid
Manual handling	Overexertion/ repetitive movement	Sprains, strains, fractures
Falls	Falls from heights, slips and trips on uneven surfaces	Fractures, bruises, cuts, dislocations, concussions
Electricity	Coming into contact with electrical potential	Shocks, burns, loss of consciousness
Plants	Hit by projectiles, striking objects, caught in machinery, overturning vehicles	Cardiac arrest, cuts, bruises, dislocations, dermatitis, fractures, amputations, eye damage
Hazardous substances	Exposure to chemicals such as solvents, acids, and hydrocarbons	Dizziness, vomiting, burns to skin and eye, respiratory problems,
Temperature, UV radiation	Effects of heat or cold from weather or work conditions	Sunburn, frostbite, heat stress, heat stroke, hypothermia
Biological	Intimidation, conflict, physical assault	Nausea, collapse, shock, physical injuries

13.2 Casualty

Casualty refers to persons involved in a serious accident, especially one involving bodily injury or death.

How to approach and care for a casualty:

1. Summon or shout for help and call emergency services.
2. Observe the area carefully while approaching the injured person to prevent yourself from being injured by the same or other hazards.
3. Calm and reassure the casualty.
4. Identify the type(s) of injury or injuries and look for signs of bleeding.
5. Identify cause(s) of injuries.
6. Check for the position of limbs.
7. Do not move or remove the injured person unless it is necessary to prevent further injuries.
8. If the casualty is conscious, speak with him or her until professional help arrives.

13.3 Injuries and Their Treatment

There are numerous injuries that may be inflicted on the job or work site. Injuries may be either lacerations or incise and include wounds, minor cuts or scrapes, deep cuts, injuries from electrical shocks, burns from fire and chemical among others.

Industrial injuries which may lead to trauma may be caused from accidents which includes falls and blows from objects. In the industrial setting, these may be prevented by:

1. Adhering to occupational and safety rules
2. Avoiding careless movements around the workplace
3. Wearing safety harness while working at heights above 6 feet
4. Wearing helmets where required

Severe industrial injuries may lead to the need to apply *cardiopulmonary resuscitation (CPR) treatment* which will be introduced later in this chapter.

13.3.1 Wounds

A wound is a break in the skin, which allows bloodletting and may allow infection to enter the body. Wounds are common among all tasks and chores; understanding the classification of wounds will ease the fearful approach to an injured person after an accident.

3.3.1.1 Classification of Wounds

The classification of wounds can be abbreviated as CLIPA.

C—Contusion
(A contusion is a bruise with bleeding under the skin. It may indicate an internal inquiry.)

L—Laceration
(A laceration is a wound caused by the tearing of soft tissue and usually causes slight bleeding. A high risk of infection or contamination accompanies them.)

I—Incision
(Incised wounds are caused by a sharp instrument or objects and usually bleed freely. There is a heightened risk of damage to underlying structures.)

P—Puncture
(A puncture wound is deceptive; it may be difficult to determine the extent of the puncture. There is a high risk of infection or internal injury.)

A—Abrasion
An abrasion will often have grit embedded. There is a high risk of infection.

13.3.2 Minor Cuts or Scrapes

Minor cuts or scrapes are surface or topical cuts that do not require complex treatment.

Treatment for minor cuts

1. Clean wound with water
2. Avoid using soap to clean wound
3. Remove dirt/debris from the wound
4. Apply gentle pressure to stop bleeding if bleeding persists
5. Apply antibiotic ointment to prevent infection
6. Dress/bandage the wound
7. Change dressing daily

13.3.3 Deep Cuts

Deep cuts are wounds which may expose underlying tissue which may require stitching and may require consultation with a doctor.

Treatment for Deep Cuts:

To effectively treat deep cuts, the following should be done:

1. Clean the wound
2. Remove deeply lodged debris
3. Breathe on the open wound
4. Push back exposed body parts
5. Seek professional medical attention.

13.3.4 Electrical Shock Injuries

Injuries from electrical shocks can be considered the most common form of injury in the electrical field. It is therefore imperative that electrical practitioners pay close attention to this section of the text. Injuries from electrical shocks may result from:

1. Electric current exceeding 4 mA passing through the body.
2. Sources which may be natural or man-made.

13.3.4.1 Possible Effects of an Electric Shock

1. Muscle contraction
2. Seizures
3. Burns
4. Makes the individual fall
5. Fractures from falling
6. Clotting of blood
7. Tissue death (narcosis)
8. Respiratory/heart/kidney failure

13.3.4.2 Treatment for Electrical Shock Victims

1. Do not attempt to move the victim from the current source with your bare hands (You may be electrocuted by doing so).
2. First step is to switch off the current source.
3. Otherwise, move the victim from the current source using a dry wooden stick.
4. Check the victim for breathing.
5. If there is no breathing, conduct CPR.
6. Call emergency medical aid.
7. If breathing, do a physical examination.
8. Treat minor burns (See section on treating burns).
9. Re-establish vital functions.
10. Excessive burns may require hospitalization/surgery.
11. Make arrangement for providing supportive care.

13.3.4.3 Prevention of Electrical Shocks

Prevention of electrical shocks lies primarily in precautions to be observed when interfacing with electricity. It is therefore imperative that the following be observed:

1. Ensure proper design, installation, and maintenance of electrical devices
2. Treat all wires as live

13.3.5 Burn Injuries

All injuries resulting from exposure to or contact with excessive heat, electricity, radiation or chemicals are referred to as burn injuries. Burn injuries are classified as follows:

1. Common heat injuries due to fire, hot liquids, or steam
2. Burns due to heat/chemicals—through skin contact
3. Severe burns which affect muscles, fat, and bones

Burn injuries are placed in three categories based on the severity of tissue damage, namely first, second and third degree.

Chemical burns are addressed separately in the next section because of its seriousness and complexity.

13.3.5.1 First-degree Burns

First-degree burn injuries are classified as mild. This burn is usually associated with swelling and redness of the injured area. The burned area will become white when touched.

Treatment of First-degree Burns

1. Carefully and quickly remove the victim from heat source.
2. Remove the burnt clothing.
3. Run cold water over the burnt area.

4. Gently clean and dry the injured area.
5. Apply antibiotic such as Silver Sulfadiazine.
6. Use a sterile bandage to cover the injured area.

13.3.5.2 Second-degree Burns

Second-degree burns extend to middle skin layer, the dermis and are accompanied by swelling, redness, blisters and pain. Like first-degree burns, the injured area may turn white when touched. Blisters may ooze a clear fluid and scars may develop when the burnt area is healed.

Treatment of Second-degree Burns

1. Clean and dry the affected area thoroughly.
2. Apply antibiotic cream over the affected area.
3. Make the patient lie down.
4. Keep burnt body part at a raised level.
5. Hospitalization may be essential.

13.3.5.3 Third-degree Burns

Third-degree burns result in damage to all 3 skin layers. It destroys adjacent hair follicles, sweat glands, and nerve endings. No blisters are usually observed, however swelling will take place in the injured area. The burnt skin usually develops a leathery texture and discoloration of the skin will take place. When healed, permanent scars will be visible, surfaces become crusty and blood circulation may be impaired in the affected area.

Treatment of Third-degree Burns

1. Requires immediate hospital care.
2. Periodically run clean, cold water over burns.
3. Nutritious diet will be required to aid speedy healing.

13.3.6 Chemical Burns

Chemical burns results from contact with strong acids or bases of alkali and chemical Irritants. Symptoms of chemical burns include the following:

1. Irritation/burning sensation
2. Redness of skin
3. Pain/numbness
4. Blisters
5. Coughing
6. Breathlessness
7. Vision loss, if the eye is affected
8. Headache

In severe cases, symptoms of chemical burns may include:

1. Dizziness
2. Severe cough
3. Seizures
4. Irregular heart beats
5. Cardiac arrest

Treatment of Chemical Burns

1. Remove the patient from the accident site.
2. Wash injury with tepid water liberally.
3. Identify chemical for effective therapy.
4. Seek medical treatment.

Prevention of Chemical Burns

1. While using chemicals, follow all safety precautions.
2. Avoid overexposure to chemicals.
3. Store chemicals safely in properly labeled containers.
4. Avoid mixing different chemicals.

13.4 Trauma

Trauma, according to the McGraw-Hill Science and Technology Encyclopaedia is Injury to tissue by physical or chemical means. Trauma to bones and joints results in fractures, dislocations, and sprains. Head injuries are often serious because of the complications of haemorrhage, skull fracture, or concussion.

13.4.1 Trauma from Cuts and Bruises

Trauma resulting from cuts and bruises is most common in the world of work and may result in the following:

1. Breaking of small veins and capillaries under the skin.
2. Blood escaping and collecting under the skin
3. Bruises looking purplish or red

Treating Trauma from Cuts and Bruises

1. Keeping bruised area raised
2. Applying a cold compress, ½ hr to one hr at a time
3. Avoiding applying ice directly on skin
4. Taking pain-killing medications like acetaminophen
5. Avoiding draining the bruise using needle

13.4.2 Head Trauma

Head trauma is an injury that affects the brain/skull. As mentioned earlier, head trauma are often serious because of the complications of haemorrhage, skull fracture, or concussion.

These injuries range from minor to serious and may be closed or penetrating. A closed head trauma results when the head hits against a blunt object, whereas a penetrating head trauma results when an object penetrates the skull and enters brain.

Head traumas are caused primarily from falls and accidents in the electrical field. There are several symptoms of head trauma which includes but not limited to:

1. Loss of consciousness—for short or long duration
2. Bleeding
3. Vomiting
4. Fluid discharge from nose
5. Loss of hearing, vision, taste, and smell
6. Speech-related problems
7. Irregular heart beats
8. Seizures
9. Paralysis

Treating Trauma from Head Injuries

1. Check for a clear airway
2. Check for breathing and start *CPR* if you are trained in administering CPR.
3. Stabilize the neck and spine, for example, using a neck brace.
4. Stop any severe bleeding.
5. Provide pain relief if required.
6. Splint any fractured or broken bones (strapping them into the correct position).

13.5 Fainting

Fainting, "blacking out," or syncope is the temporary loss of consciousness followed by the return to full wakefulness. This loss of consciousness may be accompanied by loss of muscle tone that can result in falling or slumping over. (www.MedicineNet.com)

The brain requires blood flow to provide oxygen and glucose (sugar) to its cells to sustain life. Fainting or syncope occurs when the brain is deprived of blood, oxygen, or glucose. Fainting is not caused by head trauma, since loss of consciousness after a head injury is considered a concussion. However, fainting can cause injury if the person falls and hurts themselves, or if the faint occurs while participating in an activity like driving a car.

Common Causes of Fainting include:

1. Anxiety
2. Emotional upsets
3. Stress
4. Severe pain
5. Skipping meals
6. Standing up too fast
7. Standing for a long time in a crowd
8. Diabetes
9. Blood pressure

Before fainting, a person may experience symptoms such as:

1. Nausea
2. Giddiness
3. Excessive sweating
4. Dim vision
5. Rapid heartbeat or palpitations

Treatment for Faint Victims

1. Fainting is a medical emergency until proven otherwise.
2. When a person feels the symptoms of fainting, ensures the victim sits or lies down immediately.
3. If sitting, position head between knees.
4. When a person faints, position him on his back.
5. Check to see if the airways are clear.
6. Restore blood flow by loosening clothing/belts/collars.
7. Elevate feet above head level.
8. The patient should become normal within a minute. If not, seek medical help.
9. Check if the breathing/pulse is normal.
10. If the victim does not recover, conduct CPR and/or call for help.

13.6 Cardiac Arrest

Cardiac arrest, also known as sudden cardiac arrest, is the abrupt loss of heart function in a person who may or may not have diagnosed heart disease. The time and mode of death are unexpected and occurs instantly or shortly after symptoms appear (www.heart.org).

According to the American Heart Association, cardiac arrest is caused when the heart's electrical system malfunctions. Cardiac arrest may be caused by abnormal, or irregular, heart rhythms called arrhythmias. A common arrhythmia in cardiac arrest is ventricular fibrillation (VF). This occurs when the lower chambers of the heart suddenly start beating chaotically and don't pump blood. Death will occur within minutes after the heart stops.

Treatment for cardiac arrest victims

Cardiac arrest may be reversed if cardiopulmonary resuscitation (CPR) is performed and a defibrillator is used to shock the heart and restore a normal heart rhythm within a few minutes.

1. Apply CPR to reverse VF
2. Apply a shock to the heart, using a defibrillator (Defibrillation stops VF and restores normal heart rhythm within a few minutes)

13.7 Heart Attack

The term "heart attack" is often mistakenly used to describe cardiac arrest. While a heart attack may cause cardiac arrest and sudden death, the terms don't mean the same thing. Heart attacks are caused by a blockage that stops blood flow to the heart. A heart attack (or myocardial infarction) refers to death of heart muscle tissue due to the loss of blood supply, not necessarily resulting in the death of the victim. If blood flow isn't restored quickly, the section of heart muscle begins to die.

Heart attacks most often occur as a result of coronary heart disease (CHD), also called coronary artery disease. CHD is a condition in which a waxy substance called plaque builds up inside the coronary arteries (atherosclerosis) blocking supply oxygen-rich blood to your heart. The buildup of plaque occurs over many years.

Symptoms of a heart attack are:

- **Chest pain or discomfort**. Most heart attacks involve discomfort in the center or left side of the chest.
- **Upper body discomfort**. You may feel pain or discomfort in one or both arms, the back, shoulders, neck, jaw, or upper part of the stomach.
- **Shortness of breath**. This may occur before or along with chest pain or discomfort. It can occur when you are resting or doing physical activities.
- A feeling similar to heartburn or indigestion.
- Breaking out in a cold sweat.
- Feeling unusually tired for no reason, sometimes for several days.
- Nausea (feeling sick to the stomach) and vomiting.
- Light-headedness or sudden dizziness.

Treatment for Heart Attack Victims

1. Get the victim to relax.
2. Loosen tight clothes.
3. Call for a doctor or paramedics.
4. Apply CPR or call for help if you are not trained to perform CPR.
5. 15 pumps are followed by 2 artificial respirations.
6. Continue until ambulance or doctor arrives.
7. Take a nitroglycerin pill if your doctor has prescribed this type of treatment.

13.8 Cardiopulmonary Resuscitation (CPR)

Cardiopulmonary resuscitation (CPR) is an emergency life saving measure which consists of the use of chest compressions and artificial ventilation to maintain circulatory flow and oxygenation during cardiac arrest. CPR is done on:

1. Unconscious/non breathing patients
2. Persons suffering cardiac arrest
3. Near-drowning/asphyxiation/ trauma cases

13.8.1 CPR Treatment

As outlined by Bon (n.d.) CPR comprises the following 3 steps, performed in order:

1. Chest compressions
2. Airway
3. Breathing

For lay rescuers, compression-only CPR (COCPR) is recommended.

Positioning of patient for CPR is important. CPR is most easily and effectively performed by laying the patient supine on a relatively hard surface. This position allows effective compression of the sternum. Delivering CPR on a mattress or other soft surface is generally less effective. The person giving compressions should be positioned high enough above the patient to achieve sufficient leverage, so that he or she can use body weight to adequately compress the chest.

For an unconscious adult, CPR is initiated by giving 30 chest compressions. First, the patient's head should be tilted and chin lifted to open the airway, second, determine if the patient is breathing and look in the patient's mouth for foreign body that may block the airway. Third, apply chest compressions.

To effectively apply chest compressions, the provider should do the following:

1. Place the heel of one hand on the patient's sternum and the other hand on top of the first with fingers interlaced.
2. Extend the elbows and lean directly over the patient
3. Press down, compressing the chest at least 2 in and release the chest, allowing it to recoil completely. (The compression rate should be at least 100/min "Push hard and fast")
4. After 30 compressions, 2 breaths are given; however, an intubated patient should receive continuous compressions while ventilations are given 8-10 times per minute
5. Repeat the entire process is until a pulse returns or the patient is transferred to definitive care.
6. To prevent provider fatigue or injury, a new provider should intervene every 2-3 minutes (i.e., providers should swap out, giving the chest compressor a rest while another rescuer continues CPR.

Fourth, apply ventilation. If the patient is not breathing, 2 ventilations are given via the provider's mouth or a bag-valve-mask (BVM). If available, a barrier device (pocket mask or face shield) should be used.

To perform the BVM or invasive airway technique, the provider should do the following:

1. Ensure a tight seal between the mask and the patient's face.
2. Squeeze the bag with one hand for approximately 1 second, forcing at least 500 mL of air into the patient's lungs.

To perform the mouth-to-mouth technique, the provider should do the following:

1. Pinch the patient's nostrils closed to assist with an airtight seal
2. Place mouth completely over the patient's mouth
3. After 30 chest compressions, give 2 breaths (the 30:2 cycle of CPR)
4. Give each breath for approximately 1 second with enough force to make the patient's chest rise
5. Failure to observe chest rise indicates an inadequate mouth seal or airway occlusion
6. After giving the 2 breaths, resume the CPR cycle

13.7.2 Complications of CPR

Complications of CPR include fractures of the ribs or the sternum from chest compression, this however is quite uncommon. Another complication is gastric insufflation from artificial respiration using noninvasive ventilation methods (e.g., mouth-to-mouth, BVM) which can lead to vomiting which leads to further airway compromise or aspiration. Insertion of an invasive airway prevents this problem.

13.7.3 CPR Training

CPR training is for everyone. Acquisition of knowledge and skills in this area may be the only tool you have to save a loved one's life. Outlined below are the basic steps in providing this lifeline.

13.7.3.1 Take Vital Steps to Clear the Airway (See Figure 13.1, Step 1)

1. Assess if the person is conscious/breathing
2. Lay the person on his back on a hard surface
3. Using a head tilt, lifting the chin opens his airway
4. Check for breathing sound
5. If not breathing, start mouth-to-mouth breathing

13.7.3.2 *Apply Mouth- to-mouth Breathing* (See Figure 13.1, Step 2)

1. Pinch the person's nostril shut
2. Seal his mouth with your own
3. Give the first breath, lasting one second
4. Watch if chest rises
5. If it rises, give a second rescue breath
6. If it does not rise, give a head-tilt chin-lift
7. Now give a second rescue breath

13.7.3.3 Restore Circulation through Compression (See Figure 13.1, Steps 3, 4, and 5)

1. Place the heel of your palm on the patient's chest
2. Place your other hand above first
3. Keep elbows straight
4. Push down using upper body weight (compress)
5. Push hard and fast
6. Clear airway
7. After 30 compressions, give two rescue breaths
8. Continue CPR until medical help arrives

13.7.3.4 Recovery Position - After Successful CPR (See Figure 13.2, Steps 1, 2, and 3)

Place the casualty in recovery position after application of successful CPR.

13.8.1 Practical Emergency CPR Procedure Where Required

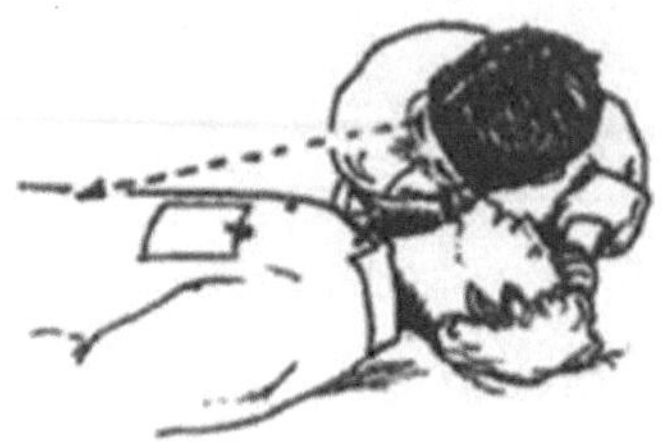

Step 1 - Tilt head and ckeck breathing

Step 2 - Close nose with two fingers and give two breaths. Chest must expand on each breath

Step 3 - Position Two fingersfrom the bottom center of chest plate

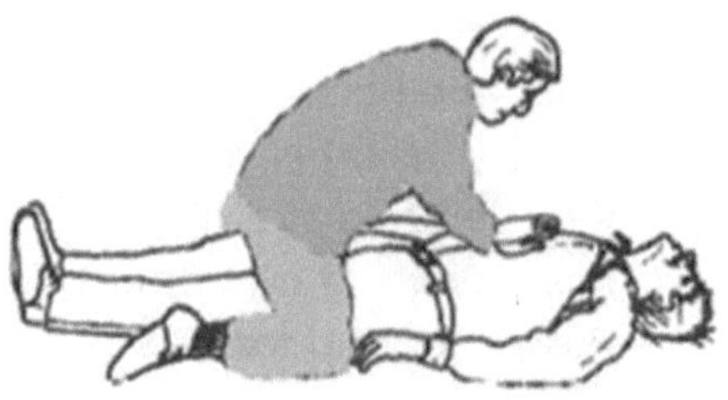

Step 4 - Push down on the 2 inches 15 times if two persons are performing CPR, and 30 times if one person is performing CPR

Step 5 - One Breath to every 15 compression if two person performing CPR, and Two Breath for 30 compression if one person is performing CPR

Figure 13.1 Performing CPR

Recovery Position—Step 1

Recovery Position—Step 2

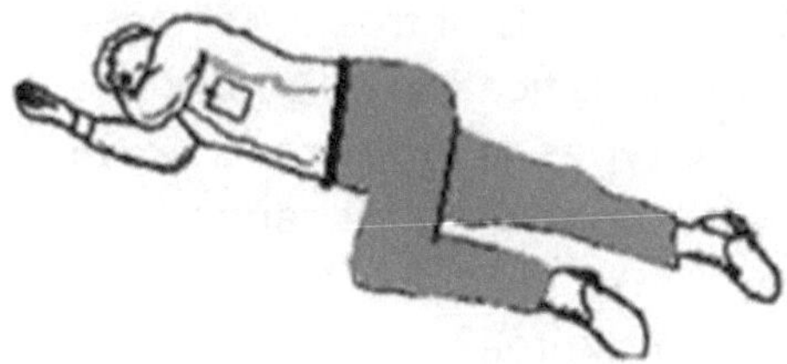

Recovery position—Step 3

Figure 13.3: *Placing casualty in recovery position*

13.9 Conclusion

First aid applications are not only essential, but critical in all aspects of life. This activity is even more critical in electrical engineering and the construction industry because of continuous exposure of persons to hazardous situations in these fields. In this chapter, some typical hazards, their typical problems, as well as the typical injuries or illnesses associated with them were presented.

A basic knowledge of the various types of injuries or illnesses, first aid techniques, and their associated applications could prevent death or casualties in the workplace. These associated principles should be kept as vital information to be used when accidents occur in order to reduce the risk of further injuries or fatalities.

13.10 Test Your Knowledge

1. Why are the basic principles of health and safety important in the workplace?
2. List the steps to be taken before attending to or while approaching a casualty.
3. If a victim was found with incise wounds, how would you care for this victim?
4. What is the meaning of abbreviation CLIPA?
5. Describe how you care for a fainted person.
6. Name two types of chemical burns.
7. Name two methods of treating chemical burns.
8. Outline the classification of burns.
9. Name three types of industrial traumas.
10. Describe the procedure for caring for victims of industrial traumas.
11. What is head trauma?
12. Explain the difference between cardiac arrest and heart attack.
13. Outline the treatment for a heart attack victim.
14. List 5 symptoms of a heart attack
15. What is the meaning of the abbreviation CPR?
16. When and why is it important to administer CPR?
17. What are the three steps that comprise CPR?
18. Describe how you would apply CPR to an unconscious person.

References

i American Heart Association (n.d.) www.heart.org

ii. Bon, Catharine A., MD (n.d.) Cardiopulmonary Resuscitation (CPR) Chief Editor: Kulkarni, Rick, MD, http://emedicine.medscape.com/article/1344081-overview

iii. heart.org (n.d.) Cardiac Arrest, http://www.heart.org/HEARTORG/Conditions/More/CardiacArrest/Cardiac-Arrest_UCM_002081_SubHomePage.jsp

iv. Medicine.net (n.d.) Fainting http://www.medicinenet.com/fainting/page2.htm

v. Medindia Health Web site (www.medindia.net) for usage of item from the following: www.medindia.net/animation/inguinal-hernia-surgery.asp

vi. Simple Safety Solutions Limited. 2002. Introduction to First Aid. Section 1 to 12.

INDEX

D

E

I

J

K

L

M

N

O

T

U

V

W

X

Y

Z

www.ingramcontent.com/pod-product-compliance
Lightning Source LLC
Chambersburg PA
CBHW020731180526
45163CB00001B/197

* 9 7 8 1 4 9 0 7 3 5 6 1 0 *